BEN GOLDACRE

I THINK YOU'LL FIND IT'S A BIT MORE COMPLICATED THAN THAT

Selected Writing

FOURTH ESTATE · *London*

Fourth Estate
An imprint of HarperCollins*Publishers*
77–85 Fulham Palace Road
London W6 8JB
www.4thestate.co.uk

First published in Great Britain in 2014 by Fourth Estate

A catalogue record for this book is available from the British Library

ISBN 978 0 00 746248 3

'Dr Goldacre Doesn't Make Everything Better' by Jeremy Laurance
© the *Independent*/www.independent.co.uk

Typeset in Minion 11/14pt by Palimpsest Book Production Limited,
Falkirk, Stirlingshire

Printed and bound in Spain by RODESA

MIX
Paper from
responsible sources

FSC **FSC C007454**
www.fsc.org

FSC™ is a non-profit international organisation established to promote
the responsible management of the world's forests. Products carrying the
FSC label are independently certified to assure consumers that they come
from forests that are managed to meet the social, economic and
ecological needs of present and future generations,
and other controlled sources.

Find out more about HarperCollins and the environment at
www.harpercollins.co.uk/green

To whom it may concern.
And Archie.
And Alice.

Contents

SURVEYS

EPIDEMIOLOGY

BAD ACADEMIA

GOVERNMENT STATISTICS

EVIDENCE-BASED POLICY

DRUGS

LIBEL

QUACKS

MAGIC BOXES

AIDS

BRAINIAC

STUFF

EARLY SNARKS

BOOKENDS

Intro

This is a collection of my most fun fights, but the fighting is just an excuse. There is nothing complicated about science, and people can understand anything, if they're sufficiently motivated. Coincidentally, people like fights. That's why I've spent the last ten years lashing science to mockery: it's the cleanest way I know to help people see the joy of statistics, and the fascinating ways that evidence can be distorted or ignored.

But these aren't personal attacks, and I'm not an angry person. All too often, people hoping to make science accessible fall into the trap of triumphalism, presenting science as a canon, and a collection of true facts. In reality, science is about the squabble. Every fight you will read in this book, over the meaning of some data, is the story of the scientific process itself: you present your idea, you present your evidence, and we all take turns to try and pull them both apart. This process of close critical appraisal isn't something we tolerate reluctantly, in science, with a grudge: far from it. Criticism and close examination of evidence is actively welcomed – it is the absolute core of the process – because ideas only exist to be pulled apart, and this is how we spiral in on the truth.

Away from the newspapers and science TV shows you can see that process, very clearly, in the institutions of science. The question-and-answer session at any academic conference, after someone presents their scientific research, is often a *bloodbath*: but nobody's resentful, everyone expects it, and we all consent to it, as a kind of intellectual S&M activity. We know it's good for our souls. If the idea survives, then great; if it needs more evidence, we decide what studies are needed next and do them. Then we all come back next year, tear the evidence apart again, and have another think. Real scientists know this. Only the fakers cry foul.

In short, this book has a manifesto: check the evidence and fight back against anyone who tries to stop you. Along the way, you will get a grounding in statistics, study design, evidence-based policy and much more, in bite-size chunks. Because while my last two books – *Bad Science* and *Bad Pharma* – were polemics with a shape, this is a racing collection of short pieces. As such, I hope it works as a kind of statistics toilet book, bringing satisfaction in short bursts, with a fight and an idea in each one.

So in the section of this book on surveys, we laugh at the stupidity of the nuclear power industry, some silly anti-abortionists and StoneWall (whom I actually adore). Or, if you prefer: we learn about the distortions of 'participant bias', misleading question design and a sticky problem involving a complex time-dependent variable. In the first piece of the book we cover some surprisingly unprofessional behaviour from a Baroness, Professor and previous Director of the Royal Institution. Or, if you prefer: we cover post-publication peer review and why the conventions of academic journals are helpful.

These pieces cover two decades of work. There are lots of *Guardian* columns, but also academic papers, a report for the UK education minister, my work in the *Romney, Hythe and Dimchurch Railway Guidebook*, the odd undergraduate essay and more. If I'm honest, it's pretty soulful (for me, not you) looking back over two decades and seeing what has changed. I was in my twenties and barely out of medical school when I started writing a column in the *Guardian*. As time passed, the targets got bigger, my day job took me through postgraduate qualifications and grown-up battles, and I think I got better at pulling claims apart. There was also discipline from outside: writing about other people's misdeeds, collecting ever greater numbers of increasingly powerful enemies – and all under British libel law – is like doing pop science with a gun to your head. So for that, thanks.

At the end I might tell you a little about how I work, why I do what I do, who made me, and how things have changed over the past two decades. For now, let's just say I'm very grateful to all the many companies and people who, by their optimistically bad behaviour under fire, have given narrative colour to what might otherwise have been some very dry explanations of basic statistical principles.

What's in this book
I've written 500,000 words in the last decade, so there is no repetition, and the corpses of folk like Gillian McKeith, the homeopaths and Big Pharma are left in my previous two books (although these characters fight on, like zombies, in the real world). My academic work on statins and Big Data is saved for a fun project that will be launching shortly. Lastly, most of my writing on randomised trials in education, policing and everywhere else is held back, as my book on this topic will come out in due course.

There is, however, some structure to this school reunion. In How SCIENCE WORKS we cover peer review, how research is unpicked and critiqued after publication, how we deal with contradictory research, the importance of methods and results being freely available, whether it matters who a researcher is, how cherry-picking harms science, and how myths are made when inconvenient results are ignored.

In BIOLOGISING we cover crass reductionism, including the peculiar beliefs that pain is only real when we scan see it on a brain scanner, that misery is best thought of as molecular, and that girls like pink 'because they evolved to look for berries'. In STATISTICS we start with easy maths and accelerate painlessly to some fairly advanced notions. We cover why the odds of three siblings sharing a birthday is not 48,627,125 to 1, why spying on us all to spot the occasional terrorist is highly unlikely to work, how statistical tools for fraud helped catch Greece faking its national economic data, what you can tell from a change in abortion rates for Down's syndrome, the many ways you can slice data to get the answer you want, the hazards of looking for spatial patterns on maps, and the most core statistical skill of all: how we can detect a true signal from everyday variation in background noise.

Then we go on to the glory of BIG DATA, the battles with government to get hold of it, the risks of sharing medical records with all and sundry, and the magical way that patterns emerge from the formless static of everyday life when you have huge numbers. In SURVEYS we learn the tricks of a sticky trade, and then we shift up a gear to cover EPIDEMIOLOGY, my day job, the science of spotting patterns in disease. Here we see how clever things called funnel plots can help to show whether one area's healthcare really is any worse than another's, whether an increase in antidepressant prescriptions really does mean more people are depressed (or even whether more people are taking antidepressants), and the core skill of all epidemiology: how to correct for 'confounding variables', or rather: how to make sure that apparent correlations in your data are real. In an overview of bicycle helmet research, we review every epidemiological error in the textbooks, and a grand claim about the benefits of screening for diseases helps show that doing something – even something small – can often be worse than doing nothing at all. We see why different study designs are needed to research common and rare diseases, and how frail memories can distort the findings, why we should never assume that laboratory tests are correlated with real patients' suffering, and how simple blinded experiments can spot if a £70 wine magnetiser really does change the flavour.

In the section on BAD ACADEMIA, we see how whole fields have been undermined by the simple misuse of statistics. We find one simple statistical error made in half of all neuroscience papers, and, by using forensic methods, we can see that brain-imaging researchers must be up to no good, because collectively they are publishing far more positive findings than the overall numbers of participants in their research could possibly, plausibly, statistically, sustain. We see bad behaviour around journals retracting papers, and appallingly poor standards in animal research, alongside academic journals publishing wildly crass papers on how, for example, people with Down's syndrome really are a bit like the Chinese.

In GOVERNMENT STATISTICS we see ludicrous over-claiming around public and private sector salaries (where commentators fail to compare like with like), Home Office figures on child abuse pulled almost from thin air, a government figure on the cost of piracy that assumes everyone in the country should be spending £9,700 each on DVDs and music every year, crime prevention numbers to support a national DNA database that simply do not add up, and a headline figure on local council overspending, from the Department for Communities and Local Government, whose derivation is so offensively stupid it almost defies belief. We also see there is no evidence that hosting events like the Olympics has any health benefit for the host nation.

EVIDENCE-BASED POLICY is a slightly different fish: is there really good evidence for the policies that governments choose? Here we see that the evidence supporting the redisorganisation of the NHS is weak and that the figures on poor performance in the NHS used to justify it are over a decade old, and when the minister tries to argue back, he digs a very deep hole. We see how a historic failure to run simple randomised trials on policy issues has left us ignorant on basic questions about what works, and then whizz through a few simple questions, showing how evidence can be checked for each one: is porn in sperm donor clinics a good idea, is organic food really better, is it wise for the Catholic Church to campaign against condoms, and are exams really getting easier? We see a thinktank report on maths, promoted by a TV maths professor, that gets its own maths catastrophically wrong, and a select committee misleading, and being misled. After all this carping, in a report for the Department For Education I set out how the teaching profession could have its own evidence-based practice revolution to mirror what we've seen in medicine (and review, along the way, how senior doctors as late as the 1970s fought back, to defend that favourite of the old and powerful: eminence-based medicine).

Recreational DRUGS are a magnet for bad policy, because ideology often conflicts with the evidence, so the temptation to distort the data is powerful. Here we see wildly inflated government figures for crop captures in Afghanistan (with a minister claiming that peasant farmers receive the entire street price from every £10 bag of heroin sold in London), and ask why death was quietly dropped from the government's measures of drug-policy success, before an essay explaining why the UK prescribed heroin for heroin addicts from the 1920s onwards, why we stopped and why we should start again.

LIBEL is a subject close to my heart, having been through the process too many times. In this section, we see how the people who sue tend not to be very nice, and how their legal aggression can – to my great pleasure – backfire. This section also includes breast-enhancement cream, and the brief return of Gillian McKeith.

I've always railed against the idea that QUACKS are manipulators, with innocent victims for customers: one woman's trip to intensive care presents an opportunity to see where the blame really lies, when quacks have their magical beliefs routinely reinforced by journalists and the government. More than that, we see how serious organisations – from universities to medicines regulators – can fail to uphold their own stated values when under political pressure or seduced by money. Then we have a brief interlude to look at three peculiarly enduring themes in modern culture: MAGIC BOXES of secret electronic components with supernatural powers (to detect bombs, cure cigarette addiction and even find murdered children), AIDS denialism (at the *Spectator*, of all places), and, in ELECTROSENSITIVITY, people eager to claim that electrical fields make you unwell (while selling you expensive equipment to protect yourself, and seducing journalists from broadsheets to the BBC's *Panorama*).

If science is about the quest for truth, then equally important is the science of IRRATIONALITY – how and why our hunches get things wrong – because that's the reason we need fair experiments and careful statistics in the first place. Here we see how our intuitions about whether a treatment works can be affected by the way the numbers are presented, how our outrage is lower when a criminal has more victims, why blind auditions can help combat sexism in orchestras, how people can turn their back on all of science when some evidence challenges just one of their prejudices, how people win more in a simple game when they're told they've got a lucky ball, how responding to a smear can reinforce it, how smokers are misled by cigarette packaging, how people can convince themselves that patients in comas are communicating, and how negative beliefs can make people experience horrible side effects, even when they're only taking sugar pills with no medicine in them. In this section I also unwisely disclose my own positive and creative visualisation ritual, and the evidence behind it.

In BAD JOURNALISM we see the many different ways that journalists can distort scientific findings: misrepresenting an MSc student's dissertation project with a headline that claims scientists are blaming women for their own rape, creating vaccine scares, and saying that exercise makes you fat. We also see the techniques journalists use to mislead, by burying the caveats and failing to link to primary sources, then we review research showing that academic press releases are often to blame, and that crass reporting on suicide can create copy-cat behaviour. The work in this section has made me extremely unpopular with whole chunks of the media, but I truly don't think there's anything personal here: the pieces are simply straight explanations, illustrating how evidence has been misrepresented by professional people with huge public influence. In light of that, I've included some attacks on me by others, and you can make what you will of their back-lash. Lastly, we see how hit TV science series *BRAINIAC* – which sells itself on doing truly dangerous, really 'real' science – simply fakes explosions with cheap stage effects.

In the final furlong, there's a collection of STUFF: my affec-tionate introduction in the guidebook of a miniature steam railway that takes you through council estates to the foot of a nuclear power station, and a guide to stalking your girlfriend through her mobile phone (with permission). Lastly there are some EARLY SNARKS. Reading your own work from ten years ago is a bit like being tied down, with your eyelids glued open, and forced to watch ten-foot videos of yourself saying stupid things with bad hair. But in case you miss the child I once was, here I take pops at cosmetics companies selling 'trionated particles', do the maths on oxygenated water that would drown you before it did any good, and cry at finding *New Scientist* being taken in by some *obviously* fake artificial intelligence software.

So welcome, again, to my epidemiology and statistics toilet book. By the simple act of keeping this book next to the loo you will – I can guarantee it – develop a clear understanding of almost all the key issues in statistics and study design. Your knowledge will outdo that of many working scientists and doctors, trapped in the silo of their specialist subjects. You will be funny at parties and useful at work, and the trionated ink molecules embedded in every page will make you youthful, beautiful and politically astute.

I hope these small packages bring you satisfaction.

2014

HOW SCIENCE WORKS

Why Won't Professor Susan Greenfield Publish This Theory in a Scientific Journal?

Guardian, 22 October 2011

This week Baroness Susan Greenfield, Professor of Pharmacology at Oxford, apparently announced that computer games are causing dementia in children. This would be very concerning scientific information; but it comes to us from the opening of a new wing at an expensive boarding school, not an academic conference. Then a spokesperson told a gaming site that's not really what she meant. But they couldn't say what she does mean.

Two months ago the same professor linked internet use with the rise in autism diagnoses (not for the first time), then pulled back when autism charities and an Oxford professor of psychology raised concerns. Similar claims go back a very long way. They seem changeable, but serious.

It's with some trepidation that anyone writes about Professor Greenfield's claims. When I raised concerns, she said I was like the epidemiologists who denied that smoking caused cancer. Other critics find themselves derided in the media as sexist. When Professor Dorothy Bishop raised concerns, Professor Greenfield responded: 'It's not really for Dorothy to comment on how I run my career.'

But I have one, humble, question: why, in over five years of appearing in the media raising these grave worries, has Professor Greenfield of Oxford University never simply published the claims in an academic paper?

A scientist with enduring concerns about a serious wide-spread risk would normally set out their concerns clearly, to other scientists, in a scientific paper, and for one simple reason. Science has authority, not because of white coats or titles, but because of precision and transparency: you explain your theory, set out your evidence, and reference the studies that support your case. Other scientists can then read it, see if you've fairly represented the evidence, and decide whether the methods of the papers you've cited really do produce results that meaningfully support your hypothesis.

Perhaps there are gaps in our knowledge? Great. The phrase 'more research is needed' has famously been banned by the *British Medical Journal*, because it's uninformative: a scientific paper is the place to clearly describe the gaps in our knowledge, and specify new experiments that might resolve these uncertainties.

But the value of a scientific publication goes beyond this simple benefit of all relevant information appearing, unambiguously, in one place. It's also a way to communicate your ideas to your scientific peers, and invite them to express an informed view.

In this regard, I don't mean peer review, the 'least-worst' system settled on for deciding whether a paper is worth publishing, where other academics decide if it's accurate, novel, and so on. This is often represented as some kind of policing system for truth, but in reality some dreadful nonsense gets published, and mercifully so: shaky material of some small value can be published into the buyer-beware professional literature of academic science; then the academic readers of this literature, who are trained to critically appraise a scientific case, can make their own judgement.

And it is this second stage of review by your peers – after

publication – that is so important in science. If there are flaws in your case, responses can be written, as letters to the academic journal, or even whole new papers. If there is merit in your work, then new ideas and research will be triggered. That is the real process of science.

If a scientist sidesteps their scientific peers, and chooses to take an apparently changeable, frightening and technical scientific case directly to the public, then that is a deliberate decision, and one that can't realistically go unnoticed. The lay public might find your case superficially appealing, but they may not be fully able to judge the merits of all your technical evidence.

I think these serious scientific concerns belong, at least once, in a clear scientific paper. I don't see how this suggestion is inappropriate, or impudent, and in all seriousness, I can't see an argument against it. I hope it won't elicit an accusation of sexism, or of participation in a cover-up. I hope that it will simply result in an Oxford science professor writing a scientific paper, about a scientific claim of great public health importance, that she has made repeatedly – but confusingly – for at least half a decade.

Cherry-Picking Is Bad. At Least Warn Us When You Do It

Guardian, 24 September 2011

Last week the *Daily Mail* and Radio 4's *Today* programme took some bait from Aric Sigman, an author of popular-sciencey books about the merits of traditional values. 'Sending babies and toddlers to daycare could do untold damage to the development of their brains and their future health,' explained the *Mail*.

These news stories were based on a scientific paper by Sigman in the *Biologist*. It misrepresents individual studies, as Professor Dorothy Bishop demonstrated almost immediately, and it cherry-picks the scientific literature, selectively referencing only the studies that support Sigman's view. Normally this charge of cherry-picking would take a column of effort to prove, but this time Sigman himself admits it, frankly, in a PDF posted on his own website.

Let me explain why this behaviour is a problem. Nobody reading the *Biologist*, or its press release, could possibly have known that the evidence presented was deliberately incomplete. That is, in my opinion, an act of deceit by the journal; but it also illustrates one of the most important principles in science, and one of the most bafflingly recent to emerge.

Here is the paradox. In science, we design every individual experiment as cleanly as possible. In a trial comparing two pills, for example, we make sure that participants don't know which pill they're getting, so that their expectations don't change the symptoms they report. We design experiments carefully like this to exclude bias: to isolate individual factors, and ensure that the findings we get really do reflect the thing we're trying to measure.

But individual experiments are not the end of the story. There is a second, crucial process in science, which is synthesising that evidence together to create a coherent picture.

In the very recent past, this was done badly. In the 1980s, researchers such as Celia Mulrow produced damning research showing that review articles in academic journals and textbooks, which everyone had trusted, actually presented a distorted and unrepresentative view when compared with a systematic search of the academic literature. After struggling to exclude bias from every individual study, doctors and academics would then synthesise that evidence together with frightening arbitrariness.

The science of 'systematic reviews' that grew from this

research is exactly that: a science. It's a series of reproducible methods for searching information, to ensure that your evidence synthesis is as free from bias as your individual experiments. You describe not just what you found, but how you looked, which research databases you used, what search terms you typed, and so on. This apparently obvious manoeuvre has revolutionised the science of medicine.

What does that have to do with Aric Sigman, the Society of Biologists, and their journal, the *Biologist*? Well, this article was not a systematic review, the cleanest form of research summary, and it was not presented as one. But it also wasn't a reasonable summary of the research literature, and that wasn't just a function of Sigman's unconscious desire to make a case: it was entirely deliberate. A deliberately incomplete view of the literature, as I hope I've explained, isn't a neutral or marginal failure. It is exactly as bad as a deliberately flawed experiment, and to present it to readers without warning is bizarre.

Blame is not interesting, but I got in touch with the Society of Biology, as I think we're more entitled to have high expectations of them than of Sigman, who is, after all, some guy writing fun books in Brighton. They agree that what they did was wrong, that mistakes were made, and that they will do differently in future.

Here's why I don't think that's true. The last time they did exactly the same thing, not long ago, with another deliberately incomplete article from Sigman, I wrote to the journal, the editor, and the editorial board, setting out these concerns very clearly.

The *Biologist* has actively decided to continue publishing these pieces by Sigman, without warning. They get the journal huge publicity; and fair enough. I'm no policeman. But in the two-actor process of communication, until it explains to its readers that it knowingly presents cherry-picked papers without warning – and makes a public commitment to stop – it's for readers to decide whether they can trust what the journal publishes.

Being Wrong

Guardian, 15 July 2011

Morons often like to claim that their truth has been suppressed: that they are like Galileo, a noble outsider fighting the rigid and political domain of the scientific literature, which resists every challenge to orthodoxy.

Like many claims, this is something for which it's possible to gather data.

Firstly, there are individual anecdotes that demonstrate the routine humdrum of medical fact being overturned. We used to think that hormone-replacement therapy significantly reduced the risk of heart attacks, for example, because this was the finding of several large observational studies. That kind of research has important limitations: if you just grab some women who are receiving prescriptions for HRT from their doctors, and compare them to women who aren't getting HRT, you might well find that the women on HRT are healthier, but that might simply be because they were healthier to start with. Women on HRT might be richer, or more interested in their health, for example. At the time, this research represented our best guess, and that's often all you have to work with. Eventually, after decades of HRT being widely used, a large randomised trial was conducted: they took 16,000 women who were eligible for HRT, and randomly assigned them to receive either real hormones, or a dummy placebo pill. At long last we had a fair test, and after several years of treatment had passed, in 2002, the answer fell out. HRT increased the risk of having a heart attack by 29 per cent.

Were these findings suppressed? No. They were greeted eagerly, and with some horror: in fact, the finding was so concerning that the trial had to be stopped early, to avoid putting any further participants at risk, and medical practice was overturned.

Even the supposed stories of outright medical intransigence turn out, on close examination, to be pretty weak: people claim that doctors were slow to embrace *Helicobacter pylori* as the cause of gastric ulcers, when in reality it only took a decade from the first murmur of a research finding to international guidelines recommending antibiotic treatment for all patients with ulcers.

But individual stories aren't enough. This week Vinay Prasad and colleagues published a fascinating piece of research about research. They took all 212 academic papers published in the *New England Journal of Medicine* during 2009. Of those, 124 made some kind of claim about whether a treatment worked or not. Then, they set about measuring how those findings fitted into what was already known. Two reviewers assessed whether the results in each study were positive or negative, and finally – separately – they decided whether these new findings overturned previous research.

Seventy-three of the studies looked at new treatments, so there was nothing to overturn. But the remaining fifty-one were very interesting, because they were, essentially, evenly split: sixteen upheld a current practice as beneficial; nineteen were inconclusive; and, crucially, sixteen found that a practice believed to be effective was in fact ineffective, or vice versa.

Is this unexpected? Not at all. If you like, you can look at the same problem from the opposite end of the telescope. In 2005, John Ioannidis gathered together all the major clinical research papers published in three prominent medical journals between 1990 and 2003; specifically, he took the 'citation classics', the forty-nine studies that were cited more than a thousand times by subsequent academic papers.

Then he checked to see whether their findings had stood the test of time, by conducting a systematic search in the literature to make sure he was consistent in finding subsequent data. Of his forty-nine citation classics, forty-five had found that an intervention was effective, but in the time that had passed subsequently, only half of these findings had been positively replicated. Seven studies – 16 per cent of the total – were flatly contradicted by subsequent research; and for a further seven studies, follow-up research had found that the benefits originally identified were present, but more modest than first thought.

This looks like a reasonably healthy state of affairs to me: there probably are true tales of dodgy peer reviewers delaying publication of findings they don't like, but overall, things are routinely proven to be wrong in academic journals. Equally, the other side of this coin is not to be neglected: we often turn out to be wrong, even with giant, classic papers. So it pays to be cautious with dramatic new findings; if you blink you might miss a refutation; and there's never an excuse to stop monitoring outcomes.

Kids Who Spot Bullshit, and the Adults Who Get Upset About It

Guardian, 28 May 2011

If you can tear yourself away from Ryan Giggs's penis for just one moment, I have a different censorship story.

Brain Gym is a schools programme I've been writing about since 2003. It's a series of elaborate physical movements with silly pseudoscientific justifications: you wiggle your head back and forth, because that gets more blood into your frontal lobes for clearer thinking; you contort your fingers together to improve some unnamed 'energy flow'. They're keen on drinking water, because 'processed foods' – I'm quoting the Brain Gym Teacher's Manual – 'do not contain water'. You pay hundreds of thousands of pounds for Brain Gym, and it's still done in hundreds of state schools across the UK.

This week I got an email from a science teacher about a thirteen-year-old pupil. Both have to remain anonymous. This pupil wrote an article about Brain Gym for her school paper, explaining why it's nonsense: the essay is respectful, straightforward, and factual. But the school decided this article couldn't be printed, because it would offend the teachers in the junior school who use Brain Gym.

Now, this is weak-minded, and perhaps even vicious. More interesting, though, is how often children are able to spot bullshit, and how often adults want to shut them up.

Emily Rosa is the youngest person ever to have published a scientific paper in the *Journal of the American Medical Association*, one of the most influential medical journals in the world. At the age of nine she saw a TV programme about nurses who practise 'Therapeutic Touch', claiming they can detect and manipulate a 'human energy field' by hovering their hands above a patient.

For her school science-fair project, Rosa conceived and executed an experiment to test if they really could detect this 'field'. Twenty-one experienced practitioners put their palms on a table, behind a screen. Rosa flipped a coin, hovered her hand over the therapist's left or right palm accordingly, and waited for them to say which it was: the therapists performed no better than chance, and with 280 attempts there was sufficient statistical power to show that these claims were bunk.

Therapeutic Touch practitioners, including some in university posts, were deeply unhappy: they insisted loudly that *JAMA* was wrong to publish the study.

Closer to home is Rhys Morgan, a schoolboy with Crohns Disease. Last year, chatting on www.crohnsforum.com, he saw people recommending 'Miracle Mineral Solution', which turned out to be industrial bleach, sold with a dreary conspiracy theory to cure Aids, cancer, and so on.

At the age of fifteen, he was perfectly capable of exploring the evidence, finding official documents, and explaining why it was dangerous. The adults banned him. Since then he's got his story on *The One Show*, while the Chief Medical Officer for Wales, the Food Standards Agency and Trading Standards have waded in to support him.

People wring their hands over how to make science relevant and accessible, but newspapers hand us one answer on a plate every week, with their barrage of claims on what's good for you or bad for you: it's evidence-based medicine. If every school taught the basics – randomised trials, blinding, cohort studies, and why systematic reviews are better than cherry-picking your evidence – it would help everyone navigate the world, and learn some of the most important ideas in the whole of science.

But even before that happens, we can feel optimistic. Information is more easily accessible now than ever before, and smart, motivated people can sidestep traditional routes to obtain knowledge and disseminate it. A child can know more about evidence than their peers, and more than adults, and more than their own teachers; they can tell the world what they know, and they can have an impact.

So the future is bright. And if you're one of the teachers who stopped a child's essay from being published because it dared to challenge your colleagues for promoting the ludicrousness of Brain Gym, then really: shame on you.

Existential Angst About
the Bigger Picture

Guardian, 21 May 2011

Here's no surprise: beliefs which we imagine to be rational are bound up in all kinds of other stuff. Political stances, for example, correlate with various personality features. One major review in 2003 looked at thirty-eight different studies, containing data on 20,000 participants, and found that overall, political conservatism was associated with things like death anxiety, fear of threat and loss, intolerance of uncertainty, a lack of openness to experience, and a need for order, structure and closure.

Beliefs can also be modified by their immediate context. One study from 2004, for example, found that when you make people think about death ('Please briefly describe the emotions that the thought of your own death arouses in you') they are more likely to endorse an essay discussing how brilliant George W. Bush was in his response to 9/11.

A new study looks at intelligent design, the more superficially palatable form of creationism, promoted by various religious groups, which claims that life on earth is too complex to have arisen through evolution and natural selection. Intelligent design implies a reassuring universe, with a supernatural creator, and it turns out that if you make people think about death, they're less likely to approve of a Richard Dawkins essay, and more likely to rate intelligent design highly.

So that's settled: existential angst drives us into the hands of religion. Rather excellently, the effect was partially reversed when people also read a Carl Sagan essay about how great it is to find meaning in the universe for yourself using science. It's perfect. I love this stuff. I love social science research that reinforces my prejudices. Everybody does.

But that's where I start to fall down. If I like these results, then lots of other people will like them too, whether it's the academic psychologists doing the research, the statisticians they collaborate with, the academic journal editors and reviewers who decide whether or not the paper gets an easy ride into print, the press officers who decide whether or not to shepherd its findings towards the public, or even, finally, the bloggers and journalists who write about it. At every step, there is room for fun results to get through, and for unwelcome results to fall off the radar.

This isn't a criticism of any individual study. Rather, it's the angst-inducing context that surrounds every piece of academic research that you read: a paper can be perfect, brilliantly well-conducted, and yet there's no way of knowing how many negative findings go missing. For all we know, we're just seeing the lucky times the coin landed heads up.

The scale of the academic universe is dizzying, after all. Our most recent estimate is that there are over 24,000 academic journals in existence, 1.3 million academic papers published every year, and over 50 million papers have been published since scholarship began.

And for every one of these 50 million papers there will be unknowable quantities of blind alleys, abandoned experiments, conference presentations, work in progress seminars, and more. Look at the vast number of undergraduate and masters dissertations that had an interesting finding, and got turned into finished academic papers; and then think about the even vaster number that didn't.

In medicine, where the stakes are tangible, systems have grown up to try to cope with this problem: trials are supposed to be registered before they begin, so we can notice the results that get left unpublished. But the systems are imperfect, and pre-registration is very rarely done, even in medical research, for anything other than trials.

We are living in the age of information, and vast tracts of

data are being generated around the world, on every continent and on every question. A £200 laptop will let you run endless statistical analyses. The most interesting questions aren't around individual nuggets of data, but rather how we can corral it to create an information architecture which serves up the whole picture.

The Glorious Mess of Real Scientific Results

Guardian, 6 November 2010

Popular science is often triumphalist, presenting research as a set of completed answers, when in reality much of what gets published makes a glorious, necessary mess.

Here is an example. Solomon Asch's legendary studies from the 1950s on conformity are among my favourite experiments of all time. Some people in a room are asked to judge the length of a line; all but one are stooges, and they unanimously assert what is obviously an incorrect answer. The one true, unsuspecting experimental subject conforms to the majority view, despite knowing that it's incorrect, about a third of the time.

This is a chilling result that feels just right, and over the past half-century researchers have replicated the study over a hundred times in seventeen countries, allowing hints of patterns to be spotted in the results. One analysis of US studies found that conformity has declined since the 1950s. Another found that 'collectivist' countries tend to show higher levels of conformity than individualist ones.

This month the *International Journal of Psychology* published a new variant. Instead of one real subject in a room full of stranger stooges, they used polarising glasses – the same technology used to present a different image to the left and right eye for 3D films – to show participants different images on the same screen, at the same time, in the same room. This meant that friends could disagree, legitimately, and so exert social pressure, but without faking it.

The results were problematic. Overall, sometimes the minority people did conform to peer pressure, giving incorrect answers. But when the results were broken down, women did conform, a third of the time, but men did not. This poses a problem. Why were the results of this study different from the original study?

It could be that the subjects were different. The Asch experiments were only conducted in men, and they did conform. Perhaps modern Japanese undergraduates are different from 1950s US undergraduates (although cultural and generational differences have not previously been shown to be so large that they abolish the conformity effect completely).

It could be that the original task, where subjects had to judge the length of a line, was slightly different. But if anything, the task in the new experiment was harder than in the original, because the polarising glasses required that extra visual noise be added in; and if judgements were trickier, and therefore closer calls, then you might expect that conformity would increase, rather than decrease.

Or it could be that the relationships were different. Perhaps conforming effects are less pronounced among people who know each other compared to an experiment with a room full of stranger stooges: perhaps you feel more comfortable disagreeing with friends. This would be an important answer, if true, because when we extrapolate from the lab to the everyday, we're probably more interested in conformity effects among acquaintances, because that's what happens in a real community.

Maybe these questions will be resolved with a new experiment – you could probably design one yourself that would discriminate between the different possible explanations – but that will depend on whether someone is interested enough, and whether they can get the money and the time. Perhaps the paper will sink like a stone, and be ignored or overlooked, as sometimes happens with uncomfortable data.

But what you should know is this: alongside the triumphalism, and the answers, in reality, grey and conflicting results like these run deep in the research literature. They're not an aberration, or a disappointment; in fact they are arguably the glorious norm, in the noise of over 20,000 academic journals, publishing well over a million articles every year. Alongside the giants, and the clean easy answers, challenging and ambiguous findings like these are what science is really made of.

Nullius in Verba*

Not in the *Guardian*, 26 June 2010

Here is some pedantry: I worry about data being published in newspapers rather than academic journals, even when I agree with its conclusions. Much like Bruce Forsyth, the Royal Society has a catchphrase: *Nullius in verba*, or 'On the word of nobody'. Science isn't about assertions on what is right, handed down from authority figures. It's about clear descriptions of studies, and the results that came from them, followed by an explanation of why they support or refute a given idea.

---

* This column uses the example of some work by the *Guardian*'s health correspondent to illustrate the importance of transparency about research methods, as well as results. This is a growing issue, as raw data and the tools for analysis have become more accessible (excitingly) – and more widely used – outside of traditional academia. It is the only column by me that the *Guardian* has ever declined to publish.

Last week the *Guardian* ran a major series of articles on the mortality rates after planned abdominal aortic aneurysm repair in different hospitals. Like many previously published academic studies on the same question, they discovered that hospitals which perform the operation less frequently have poorer outcomes. I think this is a valid finding.

The *Guardian* pieces aimed to provide new information, in that they did not use the Hospital Episodes Statistics, which have been used for much previous work on the topic (and on the NHS Choices website, where they are used to rate hospitals for the public). Instead they approached each hospital with a Freedom of Information Act request, asking the surgeons themselves for the figures on how many operations they performed, and how many people died.

Many straightforward academic papers are built out of this kind of investigative journalism work, from early epidemiology research into occupational hazards, through to the famous recent study hunting down all the missing trials of SSRI antidepressants that companies had hidden away. It's not clear whether this FOI data will be more reliable than the Hospital Episodes numbers – 'Discuss the strengths and weaknesses of the HES dataset' is a standard public health exam question – and reliability will probably vary from hospital to hospital. One unit, for example, reported a single death after ninety-five emergency AAA operations on FOI request, when on average about one in three people in the UK die during this procedure, and that suggests to me that there may be problems in the data. But there's no doubt that this was a useful thing to do, and there's no doubt that hospitals should be helpful and share this information.

So what's the problem? It's not the trivial errors in the piece, although they were there. The main *Guardian* article says there are ten hospitals with over 10 per cent mortality, but in the data there are only seven. It says twenty-three hospitals do over fifty operations a year, but looking at the data there are only twenty-one.

But here's what I think is interesting. This analysis was published in the *Guardian*, not an academic journal. Alongside the articles, the *Guardian* published its data, and as a long-standing campaigner for open access to data, I think this is exemplary. I downloaded it, as the *Guardian* webpage invited, and did a quick scatter plot, and a few other things. I couldn't see the pattern for greater mortality in hospitals that did the procedure infrequently. It wasn't barn door. Others had the same problem. I received a trickle of emails from readers who also couldn't find the claimed patterns (including a Professor of Stats, if that matters to you). Jon Appleby, Chief Economist on Health Policy at the King's Fund, posted on *Guardian* Comment is Free saying that he couldn't find the pattern either.

The journalists were also unable to tell me how to find the pattern. They referred me instead to Peter Holt, an academic surgeon who'd analysed the data for them. Eventually I was able to piece together a rough picture of what had been done, and after a few days, more details were posted online. It was a pretty complicated analysis, with safety plots and forest plots. I think I buy it as fair.

So why does it matter, if the conclusion is probably valid? Because science is not a black box. There is a reason why people generally publish results in academic journals instead of newspapers, and it's got little to do with 'peer review' and a lot to do with detail about methods, which tell us *how you know* if something is true. It's worrying if a new data analysis is published only in a newspaper, because the details of how the conclusions were reached are inaccessible. This is especially true if the analysis is so complicated that the journalists themselves did not know about it, and could not explain it, and this transparency is especially important if you're seeking to influence policy. The information needs to be somewhere.

Open data – people posting their data freely for all to re-analyse – is the big hip new zeitgeist, and a vitally important new idea. But I was surprised to find that the thing I've advocated for wasn't enough: open data is sometimes no use, unless we also have open methods.

Is It OK to Ignore Results from People You Don't Trust?

Guardian, 6 March 2010

If the media were actuarial about drawing our attention to the causes of avoidable death, your newspapers would be filled with diarrhoea, Aids and cigarettes every day. In reality we know this is an absurd idea. For those interested in the scale of our fascination with rarity, one piece of research looked at a three-month period in 2002 and found that 8,571 people had to die from smoking to generate one story on the subject from the BBC, while there were three stories for every death from vCJD.

So you've probably heard that smoking might prevent Alzheimer's. It comes up in the papers, sometimes to say it's a true finding, sometimes to say it's been refuted. Maybe you think it's a mixed bag, that the research is contradictory, that 'experts are divided'. Perhaps you smoke, and joke about how it'll stop you losing your marbles, at least.

This month, Janine Cataldo and colleagues publish a systematic review on the subject, but with a very interesting twist. First, they found all the papers ever published on smoking and Alzheimer's, using an explicit search strategy which they describe properly in the paper – because they're scientists, not homeopaths – to make sure that they found all of the evidence, rather than just the studies they already knew about, or the ones which flattered their preconceptions.

They found forty-three in total, and overall, smoking significantly increases your risk of Alzheimer's. But they went further. Eleven of the studies were written by people with affiliations to the tobacco industry. This wasn't always declared, so to double check, the researchers searched on the University of California's Legacy Tobacco Documents Library, a vast collection of scanned material that has been gathered over decades of legal action.

If you ever want to spend a chilling afternoon living in the head of an industry whose product has been proven to kill a third of its customers, this is the place for you. 'The importance of younger adults' is a tobacco industry paper that uses financial modelling to explain the importance of recruiting teenage smokers to replace the dying older ones before it's too late, and explains that 'repeated government studies have shown less than one third of smokers start after age 18 [and] only 5 per cent of smokers start after age 24'. 'Youth cigarette – new concepts' from Marketing Innovations Inc. takes these ideas further, into cola- and apple-flavour cigarettes, because 'apples connote goodness and freshness'.

How much did it matter if the researchers had worked for the tobacco companies? A lot: the risks of Alzheimer's associated with smoking reported by these papers were on average about a third lower than in those conducted by other researchers, and they produced many papers showing cigarettes were actively protective. If you exclude the eleven papers by researchers associated with the tobacco industry, and look only at the remaining thirty-two, your chances of getting Alzheimer's as a smoker are vastly higher: for the gamblers out there, comparing a smoker against a non-smoker, the odds of getting Alzheimer's are higher by 1.72 to 1.

So does that mean we can comfortably ignore all research that comes from people who disgust us? In the 1930s, identifying toxic threats in the environment became an important feature of the Nazi project to build a master race through 'racial hygiene'. Two researchers, Schairer and Schöniger, were working on biological theories of degenerate behaviour under Professor Karl Astel, a scientist who helped organise the vile 'euthanasia' operation that murdered 200,000 mentally and physically disabled people.

In 1943 they published a well-conducted case-control study demonstrating a relationship between smoking and lung cancer, almost a decade before any other researchers elsewhere. Their paper wasn't mentioned in the classic Doll and Bradford Hill paper of 1950, and if you check in the Science Citation Index, it was referred to only four times in the 1960s, once in the 1970s, and then not again until 1988, despite providing a valuable early warning on a killer that would cause 100 million early deaths in the twentieth century. It's not obvious what you should do with evidence from untrustworthy sources, but it's always worth appraising its untrustworthiness with the best tools available.

Foreign Substances in Your Precious Bodily Fluids

Guardian, 9 February 2008

You'll find fluoride in tea, beer and fish, which might sound like a balanced diet to you. This week the Health Secretary Alan Johnson announced a major new push for putting it in all tap water, with some very grand promises, and in the face of serious opposition.

In Stanley Kubrick's film *Dr Strangelove*, General Jack D. Ripper first developed his theories about environmental poisoning and bodily fluids when he experienced impotence, fatigue, and a pervasive sense of emptiness during the physical act of love. He instantly identified the cause: a communist plot to pollute our precious bodily fluids with fluoride.

Bill Etherington MP calls it a 'poison'. Campaigners say Nazis used it to subdue people in concentration camps. According to the *Guardian*'s own (sadly departed) alternative health columnist, fluoride is 'in the same league as lead and arsenic'.

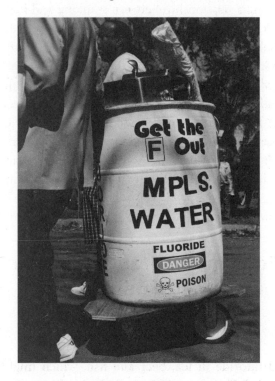

The reality is that anybody making any confident statement about fluoride – positive or negative – is speaking way beyond the evidence. In 1999 the Department of Health commissioned the Centre for Reviews and Dissemination at York University to carry out a systematic review of the evidence on the benefits of fluoridation for dental health, and to look for evidence of harm. Little new work has been done since.

They found 3,200 research papers, mostly of very poor quality. The ones which met the minimum quality threshold suggested that there was vaguely, possibly, around a 15 per cent increase in the number of children without dental caries in areas with fluoridated water, but the studies generally couldn't exclude other explanations for the variance. Of course, the big idea with fluoride in water is that it can reduce social inequalities in dental health, because everyone drinks it: but there isn't much evidence on that either – the work is of even poorer quality, and the results are inconsistent.

So when the British Dental Association says there is 'overwhelming evidence' that adding fluoride to water helps fight against tooth decay, they're with General Ripper. And when Alan Johnson says: 'Fluoridation is an effective and relatively easy way to help address health inequalities, giving children from poorer backgrounds a dental health boost that can last a lifetime,' he's really just pushing an admirably old-fashioned line that complex social problems can be addressed with £50 million worth of atoms. The people behind the York review have had to spend a fair amount of time pointing out that people are misrepresenting their work.

But since I'm in the mood for some scaremongering, let's not forget the potential harms. Fluoridation will give around one in eight people mottled teeth ('fluorosis'). And there's something else to worry about, if you like worrying. An observational study from Taiwan found a high incidence of bladder cancer in women from areas where the natural fluoride content in water was high. It might easily have been a chance finding – the study in question measured lots of variables, and if you measure enough things, then some of them are bound to come out positive, just by chance. But it could be real.

The problem here is one of small effect sizes. You don't need a careful designed study to show that falling out of a plane will probably kill you, but finding a link between fluoride and bladder cancer would be a pig to research, because the effect size is small, the exposure is spread over half a century, and the outcome – bladder cancer – takes a lifetime to reveal itself. Welcome to the finer details behind 'more research is needed'.

And the fascinating thing about public health is that, with population effects, the numbers can start to get very scary, very quickly: in the UK, for example, just a tiny 10 per cent increase in risk would give you one thousand extra new cases of bladder cancer every year. Fear. Actually, I enjoyed that. Maybe I should move to the *Mail*.

How Myths Are Made

Guardian, 8 August 2009

Much of what we cover in this column revolves around the idea of a 'systematic review', where the literature is surveyed methodically, following a predetermined protocol, to find all the evidence on a given question. As we saw in another column,* for example, the Soil Association would rather have the freedom to selectively reference only research that supports their case, rather than the totality of the evidence.

Two disturbing news stories demonstrate how this rejection of best practice can also cut to the core of academia.

Firstly, the Public Library of Science in the US this week successfully used a court order to obtain a full trail of evidence showing how pharmaceutical company Wyeth employed commercial 'ghost writers' to produce what were apparently academic review articles, published in academic journals, under

---

* See page 191.

the names of academic authors. These articles, published between 1998 and 2005, stressed the benefits of taking hormones to protect against problems like heart disease, dementia and ageing skin, while playing down the risks. Stories like this, sadly, are commonplace; but to understand the full damage that these distorted reviews can do, we need to understand a little about the structure of academic knowledge.

In a formal academic paper, every claim is referenced to another academic paper: either an original research paper, describing a piece of primary research in a laboratory or on patients; or a review paper which summarises an area. This convention gives us an opportunity to study how ideas spread, and myths grow, because in theory you could trace who references what, and how, to see an entire belief system evolve from the original data. Such an analysis was published this month in the *British Medical Journal*, and it is quietly seminal.

Steven Greenberg from Harvard Medical School focused on an arbitrary hypothesis: the specifics are irrelevant to us, but his case study was the idea that a protein called β amyloid is produced in the skeletal muscle of patients who have a condition called 'inclusion body myositis'. Hundreds of papers have been written on this, with thousands of citations between them. Using network theory, Greenberg produced a map of interlocking relationships, to demonstrate who cited what.

By looking at this network of citations he could identify the intersections with the most incoming and outgoing traffic. These are the papers with the greatest 'authority' (Google uses the same principle to rank webpages in its search results). All of the ten most influential papers expressed the view that β amyloid is produced in the muscle of patients with IBM. In reality, this is not supported by the totality of the evidence. So how did this situation arise?

Firstly, we can trace how basic laboratory work was referenced.

Four lab papers did find β amyloid in IBM patients' muscle tissue, and these were among the top ten most influential papers. But looking at the whole network, there were also six very similar primary research papers, describing similar lab experiments, which are isolated from the interlocking web of citation traffic, meaning that they received no or few citations. These papers, unsurprisingly, contained data that contradicted the popular hypothesis. Crucially, no other papers refuted or critiqued this contradictory data. Instead, those publications were simply ignored.

Using the interlocking web of citations, you can see how this happened. A small number of review papers funnelled large amounts of traffic through the network, with 63 per cent of all citation paths flowing through one review paper, and 95 per cent of all citation paths flowing through just four review papers by the same research group. These papers acted like a lens, collecting and focusing citations – and scientists' attention – on the papers supporting the hypothesis, in testament to the power of a well-received review paper.

But Greenberg went beyond just documenting bias in what research was referenced in each review paper. By studying the network, in which review papers are themselves cited by future research papers, he showed how these reviews exerted influence beyond their own individual readerships, and distorted the subsequent discourse, by setting a frame around only some papers.

And by studying the citations in detail, he went further again. Some papers did cite research that contradicted the popular hypothesis, for example, but distorted it. One laboratory paper reported no β amyloid in three of five patients with IBM, and its presence in only a 'few fibres' in the remaining two patients; but three subsequent papers cited these data, saying that they 'confirmed' the hypothesis. This is an exaggeration at best, but the power of the social network theory approach is to show what happened next: over the following ten years, these three supportive citations were the root of 7,848 supportive citation paths, producing chains of false claim in the network, amplifying the distortion.

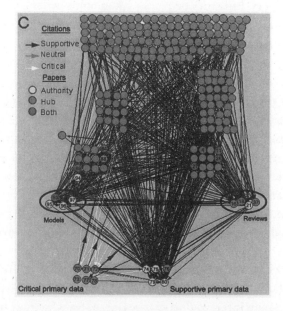

Similarly, many papers presented aspects of the β amyloid hypothesis as a theory – but gradually, through incremental mis-statement, in a chain of references, these papers came to be cited as if they proved the hypothesis as a fact, with experimental evidence, which they did not.

This is the story of how myths and misapprehensions arise. Greenberg might have found a mess, but instead he found a web of systematic and self-reinforcing distortion, resulting in the creation of a myth, ultimately retarding our understanding of a disease, and so harming patients. That's why systematic reviews are important, that's why incremental mis-statement matters, and that's why ghost writing should be stopped.

Publish or Be Damned

Guardian, 4 August 2005

I have a very long memory. So often with 'science by press release', newspapers will cover a story even though the scientific paper doesn't exist, assuming it's around the corner. In February 2004 the *Daily Mail* was saying that cod liver oil is 'nature's superdrug'. The *Independent* wrote: 'They're not yet saying it can enable you to stop a bullet or leap tall buildings, but it's not far short of that.' These glowing stories were based on a press release from Cardiff University, describing a study looking at the effect of cod liver oil on some enzymes – no idea which – that have something to do with cartilage – no idea what. I had no way of knowing whether the study was significant, valid or reliable. Nobody did, because it wasn't published. No methods, results, conclusions to appraise. Nothing.

In 1998 Dr Arpad Pusztai announced through the telly that genetically modified potatoes 'caused toxicity to rats'. Everyone was extremely interested in this research. So what had he done in his lab? What were they fed? What had he measured? A year later the paper was published, and it was significantly flawed. Nobody had been able to replicate his data and verify the supposed danger of GM, because we hadn't seen the write-up, the academic paper. How could anyone examine, let alone have a chance to rebut, Pusztai's claims? Peer review is just the start; then we have open scrutiny by the scientific community, and independent replication.

So anyway, I wrote at the time that these cod liver oil people at Cardiff University were jolly irresponsible, that patients would worry, GPs would have no answers for them, and so on. This week I contacted Cardiff and said: This is what I said last year, now where's the paper? Prof John Harwood responded through the press office: 'Mr Goldacre is quite right in asserting that scientists have to be very certain of their facts before making public statements or publishing data.' I'm a doctor, but it's good to know we agree. If puzzling.

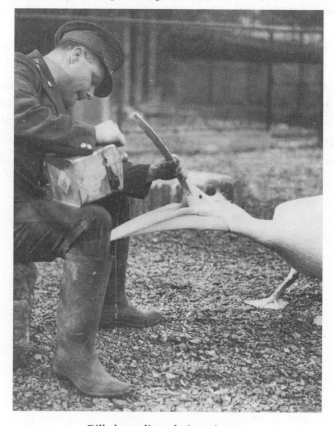

Bill the pelican being given a
dose of cod liver oil by his keeper

'Because of that,' continued Prof Harwood, 'Professor Caterson and my laboratory are continuing to work on samples.' Right . . . 'I'm afraid this takes a long time and much longer than journalists or public relations firms often realise. So, I regret he will have to be patient before Professor Caterson or myself are prepared to comment in detail.' How kind. And only slightly patronising. I don't want them to comment on fish oil. It's seventeen months after 'nature's superdrug': I just want to know where the published paper is.

In 2014, after being patient for a decade as requested, I contacted Prof Caterson and the Cardiff Press Office again. They confirmed that the research has never been published in a journal. Nobody can read or critique the methods or results, and the only public trace is a skeletal description describing a brief conference presentation. This document is four paragraphs long. The press release was seventeen paragraphs long. I'll try them again in a decade.*

---

* Where there are updates, occasionally, throughout the book, they are after the piece in italics.

Academic Papers Are Hidden from the Public. Here's Some Direct Action

Guardian, 3 September 2011

This week George Monbiot won the internet with a long *Guardian* piece on academic publishers. For those who didn't know: academics, funded mostly by the public purse, pay for the production and dissemination of academic papers; but for historical reasons, these are published by private organisations who charge around $30 per paper, keeping out any reader who doesn't have access through their institution.

This is a barrier to the public understanding of science, and also to ongoing scholarship by people who've wandered away from institutional academia. There are open-access alternatives, where academics pay up-front and the paper is free to all readers, but these are patchy, and require your funder to pay £1,000 per paper. If the journal your work is best suited for doesn't do open access, then you might reasonably accept a closed-access journal.

The arguments are big. What I find interesting is the recent rise of direct action on this issue.

Aaron Swartz is a fellow at Harvard's Center for Ethics, and a digital activist. He has been accused of intellectual property theft on a grand scale, and the federal indictment document, available in full online, describes an inspiringly nerdy game of cat and mouse.

Swartz denies all charges. Allegedly, he bought a laptop to harvest academic papers from the website JSTOR. Using a guest login at MIT – they last fourteen days – he set a program running to download papers in bulk. JSTOR and MIT smelt a rat: they blocked access to whole ranges of computers in MIT, creating havoc. Swartz set two computers on the job, running so fast that several JSTOR servers stopped working.

So then, allegedly, he tried a slower approach. You'll have seen racks of flashing network equipment in office buildings. He opened one up in a quiet basement, plugged in a laptop with some external hard drives, hid them under a box, and left this package quietly downloading papers by the million. Months later he was seen returning, peering cautiously through cracks in doors, carrying his bicycle helmet over his face and looking through the ventilation holes. He was arrested and bailed for $100,000: he had downloaded 4.8 million academic papers.

It's hard not to be impressed, and this is not the first time Swartz has taken public data access into his own hands. In the US, court records are available online, but at a cost, in a scheme generating a $150 million budget surplus. When free access was given at seventeen libraries, Swartz set up a script to harvest the lot. He got 19,856,160 pages before the system was shut down.

Now, the US government alleges that Swartz intended to release his vast academic paper stash for free on file-sharing websites. This may be true, but he did not do so. Shortly after his arrest, however, a posting appeared on the Pirate Bay website, declaring the release of an immense file, free for download. It contains thirty-three gigabytes' worth of academic papers from the UK journal *Philosophical Transactions of the Royal Society*. The release of this file, explained the poster, was an act of protest at Swartz's arrest. The papers in it range from the seventeenth century up to 1923, and are mostly out of copyright.

These are, in some respects, remarkable tales of Robin Hood behaviour. JSTOR expended huge effort on scanning these Royal Society papers in the 1990s, when scanning was tougher than it is now, and it should be thanked. But it's hard to believe we can't find any better way of allowing public access to such documents: JSTOR sells each paper for between $8 and $19, while the Royal Society estimates that the pay-per-view income from the public accessing them is half of 1 per cent of its journal income.

One major problem with the current publishing model is that it's hard to give access for free to the motivated public, while still gathering income from institutions. My hunch is that, at some stage, this problem may be partially sidestepped when someone manages an illegal workaround that individuals can play with, but which no university could endorse. I may be wrong, but either way, these are very interesting times for information.

In January 2013, facing up to thirty-five years in jail for downloading large quantities of academic papers, and under enormous pressure from US prosecutors, Aaron Swartz took his own life. He was twenty-six and extraordinary. A documentary from 2014 about Aaron's life – The Internet's Own Boy – is very good, very upsetting, and free to download online.

BIOLOGISING

Neuro-Realism

Guardian, 30 October 2010

When the BBC tells you, in a headline, that libido problems are in the brain and not in the mind, then you might find yourself wondering what the difference between the two is supposed to be, and whether a science article can really be assuming – in 2010 – that readers buy into some strange form of Cartesian dualism, in which the self is contained by a funny little spirit entity in constant pneumatic connection with the corporeal realm.

But first let's look at the experiment the BBC is reporting.

As far as we know (the study hasn't yet been published, only presented at a conference) some researchers took seven women with a 'normal' sex drive, and nineteen women diagnosed with 'hypoactive sexual desire disorder'. The participants watched a series of erotic films in a scanner, and while they did so, an MRI machine took images of blood flow in their brains. The women with a normal sex drive had an increased flow of blood to some parts of their brain – some areas associated with emotion – while those with low libido did not.

Dr Michael Diamond, one of the researchers, tells the *Mail*: 'Being able to identify physiological changes, to me provides significant evidence that it's a true disorder as opposed to a societal construct.' In the *Metro* he goes further: 'Researcher Dr Michael Diamond said the findings offer "significant evidence" that persistent low sex drive – known as hypoactive sexual desire disorder (HSDD) – is a genuine physiological disorder and not made up.'

This strikes me as an unusual world view. All mental states must have physical correlates, if you believe that the physical activity of the brain is what underlies our sensations, beliefs and experiences: so while different mental states will certainly be associated with different physical states, that doesn't tell you which caused which. If I do not have the horn, you may well fail to see any increased activity in the part of my brain that lights up when I do have the horn. That doesn't tell you why I don't have the horn: maybe I've got a lot on my plate, maybe I have a physical problem in my brain, maybe I was raped last year. There could be any number of reasons.

Far stranger is the idea that a subjective experience must be shown to have a measurable physical correlate in the brain before we can agree that the subjective experience is real, even for matters that are plainly subjective and experiential. If someone is complaining of persistent low sex drive, then they have persistent low sex drive, and even if you could find no physical correlate in the brain whatsoever, that wouldn't matter: they still have low sex drive.

Interestingly, the world view being advanced by these researchers and journalists is far from new: in fact, it's part of a whole series of recurring themes in popular misinterpretations of neuroscience, first described formally in a 2005 paper from *Nature Reviews Neuroscience* called 'fMRI in the Public Eye'. To examine how fMRI brain-imaging research was depicted in mainstream media, the researchers conducted a systematic search for every news story about it over a twelve-year period, and then conducted content analysis to identify any recurring themes.

The first theme they identified was the idea that a brain-imaging experiment 'can make a phenomenon uncritically real, objective or effective in the eyes of the public'. They described this phenomenon as 'neuro-realism', and the idea is best explained through their examples, which mirror these new claims about libido perfectly.

So, an article in the *Washington Post* takes a view on pain, and whether the subjective experience of it is enough: 'Patients have long reported that acupuncture helps relieve their pain, but scientists don't know why. Could it be an illusion?' It has an answer: 'Now brain imaging technology has indicated that the perception of pain relief is accurate.'

Another article says that brain imaging 'provides visual proof that acupuncture alleviates pain'. The reality, of course, is much simpler: for your own personal experience of pain, which is all that matters, if you say that your pain is relieved, then your pain is relieved (and I wish good luck to any doctor who tells his patient their pain has gone when it hasn't, just because some magical scan says it has).

The *New York Times* takes a similarly strange tack in a brain-imaging study on fear: 'Now scientists say the feeling is not only real, but they can show what happens in the brain to cause it.'

Many people find fatty food to be pleasurable, for the taste, the calories, and any number of other reasons. When a brain-imaging study showed that the reward centres in the brain had increased blood flow after subjects in an experiment ate high-fat foods, the *Boston Globe* explained: 'Fat really does bring pleasure.'

They're right, it does. But it's a slightly strange world when a scan of blood flow in the brain is taken as vindication of a subjective mental state, and a way to validate our experience of the world.

The Stigma Gene

Guardian, 9 October 2010

What does it mean to say that a psychological or behavioural condition has a biological cause? Over the past week, more battles have been raging over ADHD, after a paper published by a group of Cardiff researchers found evidence that there is a genetic association with the condition. Their study looked for chromosomal deletions and duplications known as 'copy number variants' (CNV), and found that these were present in 16 per cent of the children with ADHD.

What many reports did not tell you – including that in the *Guardian* – is that this same pattern of CNV was also found in 8 per cent of the children *without* ADHD. So that's not a massive difference.

But more interesting were the moral and cultural interpretations heaped onto this finding, not least by the authors themselves. 'Now we can say with confidence that ADHD is a genetic disease and that the brains of children with this condition develop differently to those of other children,' said Professor Anita Thapar. 'We hope that these findings will help overcome the stigma associated with ADHD.'

Does the belief that such problems have a biological cause really help to reduce stigma?

In 2001, Read and Harre explored attitudes among first-year undergraduate psychology students, with questionnaires designed to probe their beliefs about the causes of mental health problems, and responses on six-point scales to statements like 'I would be less likely to become romantically involved with someone if I knew they had spent time in a psychiatric hospital.' People who believed more in a biological or genetic cause were more likely to believe that people with mental health problems are unpredictable and dangerous, more likely to fear them, and more likely to avoid interacting with them. An earlier study in 1999 by Read and Law had similar results.

In 2002 Walker and Read showed young adults a video portraying a man with psychotic symptoms, such as hallucinations and delusions, then gave them either biogenetic or psychosocial explanations. Yet again, the 'medical model' approach significantly increased perceptions of dangerousness and unpredictability.

In 2004 Dietrich and colleagues conducted a huge series of structured interviews with three representative population samples in Germany, Russia and Mongolia. Endorsing biological factors as the root cause for schizophrenia was associated with a greater desire for social distance.

And lastly, more compelling than any individual study, a review of the literature to date in 2006 found that overall, biogenetic causal theories, and labelling something as an 'illness', are both positively related to perceptions of dangerousness and unpredictability, and to fear and desire for social distance. They identified nineteen studies addressing the question. Eighteen found that belief in a genetic or biological cause was associated with more negative attitudes to people with mental health problems. Just one found the opposite, that belief in a genetic or biological cause was associated with more positive attitudes.

These findings are at odds with everything that many people who campaign against stigma have assumed for many years, but they're not entirely nonsensical. As Jo Phelan explains in her paper 'Genetic Bases of Mental Illness – a Cure for Stigma?', a story about genetic causes may lead to people being conceived of as 'defective' or 'physically distinct'. It can create an 'associative stigma' for the whole family, who in turn receive new labels such as 'at risk' or 'carrier'. What's more, this stigma may persist long after the ADHD symptoms have receded in adulthood: perhaps a partner will wonder, 'Do I really want to risk having a child with this person, given their genetic predisposition?'

Perhaps it will go further than that: your children, before they even begin to show any signs of inattentiveness or hyper-activity, will experience a kind of anticipatory stigma. Do they have this condition, just like their father? 'It's genetic, you know.' Perhaps the threshold for attaining a diagnosis of ADHD will be lower for your children: it's a condition, like many others, after all, with a notably flexible diagnostic boundary.

Blaming parents is clearly vile. But before reading this research I think I also assumed, unthinkingly, like many people, that a 'biological cause' story about mental health problems was inherently valuable for combating stigma. Now I'm not so sure. People who want to combat prejudice may need to challenge their own prejudices too.

Pink, Pink, Pink, Pink. Pink Moan

Guardian, 25 August 2007

I want you to know that I love evolutionary psychologists, because their ideas – like 'Girls prefer pink because they need to be better at hunting berries' – are so much fun. Sure, there are problems. For example: we don't know a lot about life in the Pleistocene period through which humans evolved; their claims sound a bit like 'just so' stories, relying on their own internal, circular logic; the existing evidence for genetic influence on behaviour, emotion and cognition is coarse; they only pick the behaviours which they think they can explain, while leaving the rest; and they get themselves in massive trouble as soon as they go beyond examining broad categories of human behaviours across societies and cultures, becoming crassly ethnocentric. But that doesn't stop me *enjoying* their ideas.

This week every single newspaper in the world lapped up the story that scientists have cracked the pink problem. 'At last, science discovers why blue is for boys but girls really do prefer pink,' said *The Times*. And so on.

The study* took 208 people in their twenties and asked them to choose their favourite colours between two options, repeatedly, and then graphed their overall preferences. It found overlapping curves, with a significant tendency for men to prefer blue, and female subjects showing a preference for redder, pinker

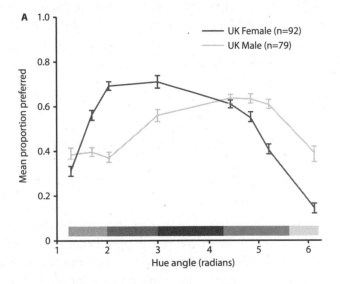

---

* Unless you have an Athens login, you are not allowed to read what the researchers actually said, instead of what the media said they said. Because although they are publicly funded academics at the University of Newcastle, and although this work has been publicised in every major mainstream media outlet in Britain and the US, and although the journal is edited by academics you fund, and paid for by subscriptions from university libraries . . . the actual academic article is behind a paywall, with a payment model geared towards institutions, rather than interested individuals. Bad luck you. I guess you have to rely on journalists.

tones. This, the authors speculated (to international excitement and approval), may be because men go out hunting, but women need to be good at interpreting flushed emotional faces, and identifying berries whilst out gathering.

Now, there are some serious problems here. Firstly, the test wasn't measuring discriminative ability, just preference. I am yet to be given evidence that my girlfriend has the upper hand in discriminating shades of red as we gambol foraging for the fruits of the forest (which we do).

But is colour preference cultural or genetic? Well. The 'girls preferring pink' thing is not set in stone, and in fact there are good reasons to suspect it is culturally determined. I have always been led to believe by my father – the toughest man in the world – that pink is the correct colour for men's shirts. In fact, until very recently blue was actively considered soft and girly, while boys wore pink, a tempered form of fierce, dramatic red.

There is no reason why you should take my word for this. Back in the days when ladies had a home journal (in 1918) the *Ladies Home Journal* wrote: 'There has been a great diversity of opinion on the subject, but the generally accepted rule is pink for the boy and blue for the girl. The reason is that pink being a more decided and stronger color is more suitable for the boy, while blue, which is more delicate and dainty, is prettier for the girl.'

The *Sunday Sentinel* in 1914 told American mothers: 'If you like the color note on the little one's garments, use pink for the boy and blue for the girl, if you are a follower of convention.' Some sources suggest it wasn't until the 1940s that the modern gender associations of girly pink became widely accepted. Pink is, therefore, perhaps not biologically girly. Boys who were raised in pink frilly dresses went down mines and fought in World War II. Clothing conventions do change over time.

But within this study, was the preference stable across cultures? Well, no, not even in this experiment, where they had some Chinese test subjects too. For these participants, not only were the differences in the overlapping curves not so extreme, but the favourite colours were a kind of red for boys and a bit pinker for girls (not blue); and they had more of a red preference overall. Red, you see, is a lucky colour in contemporary Chinese culture.

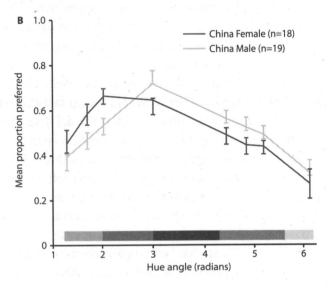

Also snuggled away in the paper was the information that femininity scores on the Bem Sex Role Inventory correlated significantly with colour preference. Now, the BSRI is a joy from the 1970s, a self-rated test explicitly designed to measure how much you adhere to socially desirable, stereotypically masculine and feminine personality characteristics.

Anyone can take this test online, for free: you simply mark on a score sheet from one to seven how much you feel you suit words like 'theatrical', 'assertive', 'sympathetic', 'adaptable' or 'tactful'; and then your score is totted up at the end. So, it turns out that women who describe themselves as 'yielding', 'cheerful', 'gullible', 'feminine' and 'do not use harsh language' also prefer pink. Thanks for the warning: I'll try to use this to avoid them in future.

It's worth being critical and thoughtful about these stories; not because it's fun to be mean, but because that's what the authors would want, and also because stories about genes and culture are an important part of the stories we tell ourselves about who and what we are, our sense of personal responsibility, and the inevitability in our gender roles.

STATISTICS

Guns Don't Kill People, Puppies Do

Guardian, 13 February 2010

Often one data point isn't enough to spot a pattern – or even to say that an event is interesting and exceptional – because numbers are all about context and constraints. At one end there are the simple examples. 'Mum Beats Odds of 50 Million-to-One to Have 3 Babies on Same Date' was the headline for the *Daily Express* on Thursday. If that phenomenon was really so unlikely, then since there are fewer than a million births a year in the UK, this would genuinely be a very rare event.

The *Express*'s number is calculated as $365 \times 365 \times 365 = 48,627,125$. But in fact it's out by an order of magnitude. One in 50 million would be the odds against someone having three siblings who share one particular *prespecified* birth date, which the editors of the *Daily Express* had sealed in an envelope and given to a lawyer fifty years ago. In reality there is no constraint on which day the *first* baby gets born on, so after that, we're just interested in the odds of two more babies sharing that same birthday, which are $365 \times 365 = 133,225$ to one. And those odds might even be a bit lower: if you two feel friskier in winter, for example, your babies might tend to be born in the autumn.

Then there is the context. Living on your street, hanging out with the people from work, it's easy to miss the sheer scale of humanity on the planet. In England and Wales there were 725,440 births last year. From the Office for National Statistics

(ONS) Statistical Bulletin 'Who is Having Babies', 14 per cent of these were third births, and another 9 per cent were fourth or subsequent births. So there are 102,000 third children born a year, 167,000 third or more-th children, and if we include the rest of the Kingdom there are even more. All of which means that on average – since the odds of three shared birthdays is about 133,225 to one, and there are 167,000 third births a year in England – this specific birthday coincidence will occur about once or twice a year in the UK. To be written about in the *Express* it would also need to be a birth within a marriage, which gives us 55,000 chances a year, or once every two years.

When you forget about numerical constraints, all kinds of things can start to look spooky: in a group of twenty-three people, there is a 50 per cent chance that two of them will share a birthday, because any pair of birthdays on any date is acceptable. When you forget about numerical context, things can look weird too: if Uri Geller gets a nation in front of the telly to tap their broken watches against the screen, and ring the call centre if the watch starts ticking again, then with viewing figures of a few million there will be more excited calls than the switchboard can handle.

If you turned to your friend and said, 'You know, a lot of funny things have happened to me, quite unexpectedly, over the course of a lifetime, but let me take a moment to specify right now the one thing that would seriously freak me out, over the next twelve hours, which would be if my dog trod on the trigger of my gun, and accidentally shot me in the face,' and then your dog shot you in the calf, that would be weird. So 'Dog Shoots Man' was a big story in America this week, to the delight of headline writers. But here's 'Dog Shoots Man in the Back' from Memphis in 2007, another in Iowa only two months later, and my own personal favourite: 'Puppy Shoots Man: Dog Put Paw on Gun's Trigger as Owner Tried to Kill Him'.

Guns don't kill people, puppies do. The world is a really big place.

Datamining for Terrorists Would Be Lovely If It Worked

Guardian, 28 February 2009

This week Sir David Omand, the former Whitehall Security and Intelligence Co-ordinator, described how the state should analyse data about individuals in order to find terrorist suspects: travel information, tax and phone records, emails, and so on. 'Finding out other people's secrets is going to involve breaking everyday moral rules,' he said, because we'll need to screen everyone to find the small number of suspects.

There is one very significant issue that will always make datamining unworkable when used to search for terrorist suspects in a general population, and that is what we might call the 'baseline problem': even with the most brilliantly accurate test imaginable, your risk of false positives increases to unworkably high levels, as the outcome you are trying predict becomes rarer in the population you are examining. This stuff is tricky but important. If you pay attention you will understand it.

Let's imagine you have an amazingly accurate test, and each time you use it on a true suspect, it will correctly identify them as such eight times out of ten (but miss them two times out of ten); and each time you use it on an innocent person, it will correctly identify them as innocent nine times out of ten, but incorrectly identify them as a suspect one time out of ten.

These numbers tell you about the chances of a test result being accurate, given the status of the individual, which you already know (and the numbers are a stable property of the test). But you stand at the other end of the telescope: you have the result of a test, and you want to use that to work out the status of the individual. That depends entirely on how many suspects there are in the population being tested.

If you have ten people, and you know that one of them is a suspect, and you assess them all with this test, then you will correctly get your one true positive and – on average – one false positive. If you have a hundred people, and you know that one is a suspect, you will get your one true positive and, on average, ten false positives. If you're looking for one suspect among a thousand people, you will get your suspect, and a hundred false positives. Once your false positives begin to dwarf your true positives, a positive result from the test becomes pretty unhelpful.

Remember this is a screening tool, for assessing dodgy behaviour, spotting dodgy patterns, in a general population. We are invited to accept that everybody's data will be surveyed and processed, because MI5 has clever algorithms to identify people who were never previously suspected. There are 60 million people in the UK, with, let's say, 10,000 true suspects. Using your unrealistically accurate imaginary screening test, you get 6 million false positives. At the same time, of your 10,000 true suspects, you miss 2,000.

If you raise the bar on any test, to increase what statisticians call the 'specificity', and thus make it less prone to false positives, then you also make it much less sensitive, so you start missing even more of your true suspects (remember, you're already missing two in ten of them).

Or do you just want an even more stupidly accurate imaginary test, without sacrificing true positives? It won't get you far. Let's say you incorrectly identify an innocent person as a suspect one time in a hundred: you still get 600,000 false positives, out of the UK population. One time in a thousand? Come on. Even with these unfeasibly accurate imaginary tests, when you screen a general population as proposed, it is hard to imagine a point where the false positives are usefully low, and the true positives are not missed. And our imaginary test really was ridiculously good: it's a very difficult job to identify suspects just from slightly abnormal patterns in the normal things that everybody does.

Then it gets worse. These suspects are undercover operatives: they're trying to hide from you, they know you're datamining, so they will go out of their way to produce trails which can confuse you.

And lastly, there's the problem of validating your algorithms, and calibrating your detection systems. To do that, you need training data: 10,000 people where you know for definite if they are suspects or not, to compare your test results against. It's hard to picture how that can be done.

I'm not saying you shouldn't spy on everyday people; obviously I have a view, but I'm happy to leave the morality and the politics to those less nerdy than me. I'm just giving you the maths on specificity, sensitivity and false positives.

Benford's Law: Using Stats to Bust an Entire Nation for Naughtiness

Guardian, 17 September 2011

This week we might bust an entire nation for handing over dodgy economic statistics. But first: why would they bother? Well, it turns out that whole countries have an interest in distorting their accounts, just like companies and individuals. If you're a Euro member like Greece, for example, you have to comply with various economic criteria, and there's the risk of sanctions if you miss them.

Government figures are subjected to various forms of audit already, of course, but alongside checking that things marry up with each other, forensic statisticians also have a few interesting tricks to try to spot suspicious patterns in the raw numbers, and so estimate the chances that figures from a set of accounts have been tampered with. One of the cleverest tools is something called Benford's law.

Imagine you have the data on, say, the population of every country in the world. Now, take only the 'leading digit' from each number: the first number in the number, if you like. For the UK population, which was 61,838,154 in 2009, that leading digit would be six. Andorra's was 85,168, so that's eight. And so on.

If you take all those leading digits, from all the countries, then overall you might naïvely expect to see the same number of ones, fours, nines and so on. But in fact, for naturally occurring data, you get more ones than twos, more twos than threes, and so on, all the way down to nine. This is Benford's law: the distribution of leading digits follows a logarithmic distribution, so you get a 'one' most commonly, appearing as first digit around 30 per cent of the time, and a nine as first digit only 5 per cent of the time.

The next time you're waiting for a bus, you can think about why this happens (bear in mind what leading digits do when quantities repeatedly double, perhaps). Reality agrees with this theory pretty neatly, and if you go to the website testingbenfordslaw.com you'll see the proportions of each leading digit from lots of real-world datasets, graphed alongside what Benford's law predicts they should be, with data ranging from Twitter users' follower counts to the number of books in different libraries across the US.

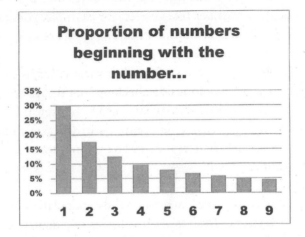

Benford's law doesn't work perfectly: it only works when you're examining groups of numbers that span several orders of magnitude. So, for example, for the age in years of the graduate working population, which goes from around twenty to seventy, it wouldn't be much good; but for personal savings, from nothing to millions, it should work fine. And of course Benford's law works in other counting systems, so if three-fingered sloths ever developed numeracy, and counted in base 6, or maybe base 12, the law would still hold.

This property of naturally occurring data has been used to check for dubious behaviour in figures for four decades now: it was first used on socioeconomic data submitted to support planning applications, and then on company accounts; it's even admissible in US courts. In 2009 an economist from the Bundesbank suggested using Benford's law on countries' economic data, and last month the results were published (hat-tip to Tim Harford for the paper).

Researchers took macroeconomic data on all twenty-seven EU nations, looking specifically at the accounting data that countries have to hand over for monitoring, which is all posted for free at the online repository Eurostat: things like government deficit, debt, revenue, expenditure, and so on. Then they took just the first digits from all the numbers, and checked to see if they deviated from what you would predict using Benford's law.

The results were fun. Greece – whose economy has tanked – showed the largest and most suspicious deviation from Benford's law of any country in the Euro.

This isn't a massive surprise: the EU has run several investigations into Greece's numbers already, and the ones from 2005 to 2008 were repeatedly revised upwards after the fact. But it's neat, and if you wanted to while away a very nerdy afternoon, could even download the data, for free from Eurostat, and repeat the analysis for yourself. Joy!

The Certainty of Chance

Guardian, 6 September 2008

Britain's happiest places have been mapped by scientists, according to the BBC: Edinburgh is the most miserable place in the country, and they were overbrimming with technical details on exactly how miserable we are in each area of Britain. The story struck a chord, and was lifted by journalists throughout the nation, as we cheerfully castigated ourselves. 'Misera-Poole?' asked the *Dorset Echo*. 'No Smiles in Donny', said *Doncaster Today*.

From the *Bromley Times*, through Bexley, Dartford and Gravesham, to the *Hampshire Chronicle*, everyone was keen to analyse and explain their ranking. 'Basingstoke lacks any sense of community or heart,' said Reverend Dr Derek Overfield, industrial chaplain for the area. And so on.

Exactly what kind of data is the good reverend explaining there? *The Times* had some methodological information: 'Researchers at Sheffield and Manchester universities based their findings on more than 5,000 responses from the annual British Household Panel Survey.' According to the BBC it was presented in a lecture at some geographical society. 'However,' it said quietly, 'the researchers stress that the variations between different places in Britain are not statistically significant.'

Here, nestled away, halfway through the gushing barrage of data and facts, was an unmarked confession: this entire news story was based on nothing more than random variation.

There are many reasons why you might see differences between different areas in your survey data on how miserable people are, and people being differently miserable is only one explanation. There might also be, of course, the play of chance: 5,000 people in 274 areas doesn't give you many in each town – fewer than twenty, in fact – so you might just happen to have picked out more miserable people in Edinburgh, and miss the fact that misery is uniformly distributed throughout the country.

This is called sampling error, and it quietly undermines almost every piece of survey data ever covered in any newspaper. Although the phenomenon has spawned a fiendish area of applied maths called 'statistics', the basic principles are best understood with a simple game.

Dr W. Edwards Deming was a charismatic American management guru who railed against performance-related pay on the grounds that it arbitrarily rewarded luck. Working in a theatrical field, he demonstrated his ideas with a simple piece of stagecraft he called the Red Bead Experiment.

Deming would appear at management conferences with a big trough containing thousands of beads which were mostly white, but 20 per cent were red. Eight volunteers were then invited up on stage from the audience of management drones: three to be managers, and five to be workers. 'Your job,' Deming explained solemnly, 'is to make white beads.'

He then produced a paddle with fifty holes cut into it, which was passed to each 'worker' in turn. They dipped the paddle into the trough, wiggled it around, and tried to produce as many white beads as they could manage through this entirely random process.

'Go and show the inspectors,' Deming would say sternly.

'Only five red beads, well done! Fourteen red beads? I think we need to re-evaluate your skill set.' Workers were sacked, promoted, retrained and redeployed, to great amusement.

We ignore basic principles like sampling error at our peril, because the illusion of control, which we all carry around for the sake of sanity, is more powerful than we think, and countless workers have had their lives turned to misery for the simple crime of pulling out fifteen red beads.

Back in the world of misery, were the journalists blameless, and guilty only of ignorance? For any individual, nobody can tell. But Dr Dimitris Ballas, the academic who did the research, has a clue: 'I tried to explain issues of significance to the journalists who interviewed me. Most,' he says, 'did not want to know.'

Sampling Error, the Unspoken Issue Behind Small Number Changes in the News

Guardian, 20 August 2011

What do all these numbers mean? '"Worrying" Jobless Rise Needs Urgent Action – Labour' was the BBC headline. It explained the problem in its own words: 'The number of people out of work rose by 38,000 to 2.49 million in the three months to June, official figures show.'

There are dozens of different ways to quantify the jobs market, and I'm not going to summarise them all here. The claimant count and the labour force survey are commonly used, and the number of hours worked is informative too: you can fight among yourselves over which is best, and get distracted by party politics to your hearts' content. But in claiming that this figure for the number of people out of work has risen, the BBC is simply wrong.

Here's why. The 'Labour Market' figures come through the Office for National Statistics, and it has published the latest numbers in a PDF document. On page 13, top table, fourth row, you will find these figures the BBC is citing. Unemployment aged sixteen and above is at 2,494,000, and has risen by 38,000 over the past quarter (and by 32,000 over the past year). But you will also see some other figures, after the symbol '±', in a column marked 'sampling variability of change'.

Those figures are called '95 per cent confidence intervals', and these are among the most useful inventions of modern life.

We can't do a full census of everyone in the population every time we want some data, because they're too expensive and time-consuming for monthly data collection. Instead, we take what we hope is a representative sample.

This can fail in two interesting ways. Firstly, you'll be familiar with the idea that a sample can be *systematically* unrepresentative: if you want to know about the health of the population as a whole, but you survey people in a GP waiting room, then you're an idiot.

But a sample can also be unrepresentative simply by chance, through something called sampling error. This is not caused by idiocy. Imagine a large bubblegum-vending machine, containing thousands of blue and yellow bubblegum balls. You know that exactly 40 per cent of those balls are yellow. When you take a sample of a hundred balls, you might get forty yellow ones, but in fact, as you intuitively know already, sometimes you will get thirty-two, sometimes forty-eight, or thirty-seven, or forty-three, or whatever. This is sampling error.

Now, normally, you're at the other end of the telescope. You take your sample of a hundred balls, but you don't know the true proportion of yellow balls in the jar – you're trying to estimate that – so you calculate a 95 per cent confidence interval around whatever proportion of yellow you get in your sample of a hundred balls, using a formula (in this case, $1.96 \times$ the square root of $((0.6 \times 0.4) \div 100)$).

What does this mean? Strictly (it still makes my head hurt), it means that if you repeatedly took samples of a hundred, then on 95 per cent of those attempts, the true proportion in the bubblegum jar would lie somewhere between the upper and lower limits of the 95 per cent confidence intervals of your samples. That's all we can say.

So, if we look at these employment figures, you can see that the changes reported are clearly not statistically significant: the estimated change over the past quarter is 38,000, but the 95 per cent confidence interval is ±87,000, running from –49,000 to 125,000. That wide range clearly includes zero, which means it's perfectly likely that there's been no change at all. The annual change is 32,000, but again, that's ±111,000.

I don't know what's happening to the economy – it's probably not great. But these specific numbers are being over-inter-preted, and there is an equally important problem arising from that, which is frankly more enduring for meaningful political engagement.

We are barraged, every day, with a vast quantity of numer-ical data, presented with absolute certainty and fetishistic preci-sion. In reality, many of these numbers amount to nothing more than statistical noise, the gentle static fuzz of random variation and sampling error, making figures drift up and down, following no pattern at all, like the changing roll of a dice. This, I confidently predict, will never change.

Scientific Proof That We Live in a Warmer and More Caring Universe

Guardian, 29 November 2008

As usual, it's not Watergate, it's just slightly irritating. 'Down's births increase in a caring Britain', said *The Times*: 'More babies are being born with Down's syndrome as parents feel increas-ingly that society is a more welcoming place for children with the condition.' That's beautiful. 'More mothers are choosing to keep their babies when diagnosed with Down's Syndrome' said the *Mail*. 'Parents appear to be more willing to bring a child with Down's syndrome into the world because British society has become increasingly accepting of the genetic abnormality' said the *Independent*. "Children's quality of life is better and acceptance has risen', said the *Mirror*.

Their quoted source was no less impeccable than a BBC Radio 4 documentary presented by Felicity Finch (her what plays Ruth Archer), broadcast on Monday. 'The number of babies with Down syndrome has steadily fallen, that is until today, when for the first time ever that number is higher than before, when testing was introduced.' I see. 'I'm keen to find out why more parents are making this decision.' They're not. 'I was so intrigued by these figures that I've been following some parents to find out what lies behind their choice.' Felicity, they're not. The entire founding premise of your entire twenty-seven-minute documentary is wrong.

There has indeed been a 4 per cent increase in Down's syndrome live births in England and Wales from 1989 to 2006 (717 and 749 affected births in the two years, respectively). However, since 1989 there has also been a far greater increase in the number of Down's syndrome foetuses created in the first place, because people are getting pregnant much later in life.

What causes Down's syndrome? We don't really know, but maternal age is the only well-recognised association. Your risk of a Down's syndrome pregnancy below the age of twenty-five is about one in 1,600. This rises to about one in 340 at thirty-five, and one in forty at the age of forty-three. In 1989, 6 per cent of pregnant women were over thirty-five years of age. By 2006 it was 15 per cent.

The National Down Syndrome Cytogenetic Register holds probably the largest single dataset on Down's syndrome, with over 17,000 anonymous records collected since 1989, making it one of the most reliable resources in the search for patterns and possible causal factors. They have calculated that if you account for the increase in the age at which women are becoming pregnant, from 1989 to 2006 the number of Down's syndrome live births in the UK would have increased not by 4 per cent, but from 717 to an estimated 1,454, if screening and subsequent termination had not been available.

Except, of course, antenatal screening is widely available, it is widely taken up, and contrary to what every newspaper told you this week, it is widely acted upon. More than nine out of ten women who have an antenatal diagnosis of Down's syndrome decide to have a termination of the pregnancy. This proportion has not changed since 1989. This is the 'decision' that Felicity Finch, Radio 4, the *Mail*, *The Times*, the *Mirror* and the rest are claiming more parents are taking: to carry on with a Down's syndrome pregnancy. This is what they are taking as evidence of a more caring society. But the figure has not changed.

Since we've now established beyond any doubt that the team behind this documentary got their numbers – and therefore their whole factual premise – entirely wrong, I think we're also entitled to engage with their crass moral judgements. If I terminate a Down's syndrome pregnancy, is that proof that society is not a warm, caring place, and that I am not a warm, caring person? For many parents the decision to terminate will be a difficult and upsetting one, especially later in life, and stories like this create a pretty challenging backdrop for making it. This would have been true even if the programme-makers had got their figures absolutely perfect, but as is so often the case for those with spare flesh to wave at strangers, their facts and figures are simply wrong.

The National Down Syndrome Cytogenetic Register felt obliged to issue a thorough clarification. The thoroughly brilliant 'Behind the Headlines' service on the NHS Choices website took the story to pieces, as it so often does, in its daily round-up of the real evidence behind the health news (disclosure: I had a trivially tiny hand in helping to set this service up).

Everybody ignored them, nobody has clarified, and *Born With Down's* remains 'Choice of the Day' on the Radio 4 website.

Drink Coffee, See Dead People

Guardian, 17 January 2009

'Danger from just 7 cups of coffee a day', said the *Express* on Wednesday. 'Too much coffee can make you hallucinate and sense dead people say sleep experts. The equivalent of just seven cups of instant coffee a day is enough to trigger the weird responses.' The story appeared in almost every national newspaper.

This was weak observational data. That's just the start of our story, but you should know exactly what the researchers did. They sent an email inviting students to fill out an online survey, and 219 agreed.

The survey is still online (I just clicked answers randomly to see the next question until I got to the end). It asks about caffeine intake in vast detail, and then uses one scale to measure how prone you are to feeling persecuted, and another, the 'Launay-Slade Hallucination Scale', sixteen questions designed to measure 'predisposition to hallucination-like experiences'.

Some of these questions are about having hallucinations and seeing ghosts, but some really are a very long way from there. Heavy coffee drinkers could have got higher scores on this scale by responding affirmatively to statements like: 'No matter how hard I try to concentrate on my work, unrelated thoughts always creep into my mind'; 'Sometimes a passing thought will seem so real that it frightens me'; or 'Sometimes my thoughts seem as real as actual events in my life.' That's not seeing ghosts or hearing voices.

And of course, this was weak observational data, and there could have been alternative explanations for the observed correlation between caffeine intake and very slightly higher LSHS scores. Maybe some students who drink a lot of coffee are also sleep-deprived, and marginally more prone to hallucinations because of that. Maybe they are drinking coffee to help them get over last night's massive marijuana hangover.

Maybe the kinds of people who take drugs instrumentally to have fun and distort their perceptions also take drugs like caffeine instrumentally to stay alert. You can think of more, I'm sure. The researchers were keen to point out this shortcoming in their paper. The *Express* and many others didn't seem to care.

Then, if you read the academic paper, you find that the associations reported are weak. For the benefit of those who understand 'regression' (and it makes anybody's head hurt), 18 per cent of the variance in the LSHS score is explained by gender, age and stress. When you add in caffeine to those three things, 21 per cent of the variance in the LSHS score is explained: only an extra 3 per cent, so caffeine adds very little. The finding is statistically significant, as the researchers point out, so it's unlikely to be due to chance, but that doesn't change the fact that it's still weak, and it still explains only a tiny amount of the overall variance in scores on the 'predisposed-to-hallucinations' scale.

Lastly, most newspapers reported a rather dramatic claim, that seven cups of coffee a day is associated with a three times higher prevalence of hallucinations. This figure does not appear anywhere in the paper. It seems to be an ad hoc analysis done afterwards by the researchers, and put into the press release, so you cannot tell how they did it, or whether they controlled appropriately for problems in the data, like something called 'multiple comparisons'.

Here is the problem. Apparently this three-times-greater risk is for the top 10 per cent of caffeine consumers, compared with the bottom 10 per cent. They say that heavy caffeine drinkers were three times more likely to have answered affirmatively to just one LSHS question: 'In the past, I have had the experience of hearing a person's voice and then found that no one was there.'

Now, this poses massive problems. Imagine that I am stood facing a barn, holding a machine gun, blindfolded, firing off shots whilst swinging my whole body from side to side and laughing maniacally. I then walk up to the barn, find three bullet holes which happen to be very close together, and draw a target around them, claiming I am an excellent shot.

You can easily find patterns in your data once it's collected. Why choose 10 per cent as your cut-off? Why not the top and bottom quarters? Maybe they have accounted for this problem. You don't know. I don't know. They say they have, to me, in emails, but it wasn't in the paper, and we can't all see the details. I don't think that's satisfactory for a headline finding, and the first claim of a press release.

Then there is one final problem: putting a finding in the press release but not into the paper is a subversion of the peer-review process. People will read this coverage, they will be scared, and they will change their behaviour. But the researchers' key reported claim, with massive popular impact, was never peer reviewed, and crucially the technical details behind it are not in the public domain.

I'm sorry to see academics unblameless in this dreary situation.

Voices of the Ancients

Guardian, 16 January 2010

Every now and then you have to salute a genius. Both the *Daily Mail* and the *Metro* report new research analysing the positions of Britain's ancient sites, and the results are startling: primitive man had his own form of 'sat nav'. Researcher Tom Brooks analysed 1,500 prehistoric monuments, and found them all to be on a grid of isosceles triangles, each pointing to the next site, allowing our ancestors to travel between settlements with pinpoint accuracy. The papers even carried an example of his map work, which I have reproduced here.

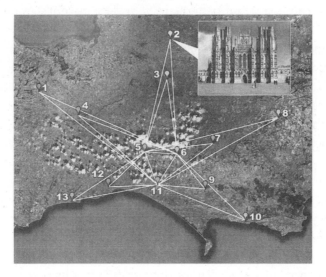

That this pattern could occur simply because one site was on the way to the next was not considered. Mr Brooks has proven, he explains, that there were keen mathematicians here 5,000 years ago, millennia before the Greeks invented geometry: 'Such is the mathematical precision, it is inconceivable that this work could have been carried out by the primitive indigenous culture we have always associated with such structures . . . all this suggests a culture existing in these islands in the past quite outside our expectation and experience today.' He does not rule out extraterrestrial help.

In the *Metro* Tom Brooks is a researcher. To the *Daily Mail* he is a researcher, a historian and a writer. I hope it's not rude or unfair for me to add 'retired marketing executive of Honiton, Devon'.

Matt Parker, his nemesis, is based in the School of Mathematical Sciences at Queen Mary, University of London. He has applied the same techniques used by Mr Brooks to another mysterious and lost civilisation.

'We know so little about the ancient Woolworths stores,' he explains, 'but we do still know their locations. I thought that if we analysed the sites we could learn more about what life was like in 2008 and how these people went about buying cheap kitchen accessories and discount CDs.'

The results revealed an exact and precise geometric placement of the Woolworths locations. 'Three stores around Birmingham formed an exact equilateral triangle (Wolverhampton, Lichfield and Birmingham stores), and if the base of the triangle is extended, it forms a 173.8-mile line linking the Conwy and Luton stores. Despite the 173.8-mile distance involved, the Conway Woolworths store is only forty feet off the exact line and the Luton site is within thirty feet. All four stores align with an accuracy of 0.05 per cent.'

Matt Parker used an ancient technique: he found his patterns in eight hundred ex-Woolworths locations by 'skipping over the vast majority, and only choosing the few that happen to line up'.

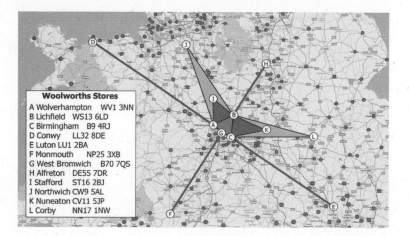

Woolworths Stores
A Wolverhampton WV1 3NN
B Lichfield WS13 6LD
C Birmingham B9 4RJ
D Conwy LL32 8DE
E Luton LU1 2BA
F Monmouth NP25 3XB
G West Bromwich B70 7QS
H Alfreton DE55 7DR
I Stafford ST16 2BJ
J Northwich CW9 5AL
K Nuneaton CV11 5JP
L Corby NN17 1NW

With 1,500 locations, Mr Brooks had almost twice as much data to work with, and on this issue Parker is clear: 'It is extremely important to look at how much data people are using to support an argument. For example, the case for global warming covers vast amounts of comprehensive evidence, but it is still possible for people to search through the data and find a few isolated examples that appear to show otherwise.'

BIG DATA

There's Something Magical About Watching Patterns Emerge from Data

Guardian, 11 June 2011

We all know that one atom of experience isn't enough to spot a pattern: but when you put lots of experiences together and process that data, you get new knowledge. This might sound obvious, but following it through – watching patterns emerge from the noise – still gives me a sense of beauty and awe.

A paper in the *British Medical Journal* this week is a perfect example. Medicine is an imperfect art, so it's inevitable that healthcare workers will make some suboptimal decisions: not so much the dramatic stuff – injecting people with the wrong drug – but more the marginal decisions, at the edges of the tweaks in a patient's journey, affecting outcomes in ways that are harder to predict.

These kinds of complex decisions will inevitably be affected by context, and one example of that context is the franticness of A&E. Waiting times are a problem in a lot of countries. In the UK we introduced a four-hour ceiling as our target, and most hospitals met it. Abolishing that four-hour target was one of the coalition government's first NHS reforms. But do waiting times matter?

Some researchers in Canada decided to find out. They gathered

73

data from all the people who visited any A&E department in Ontario over a five-year period: this gave them data on a dizzying 22 million visits. Of these, 14 million resulted in the patient being seen and then sent home. Then they followed these patients up to see what happened, and specifically, to see if they died.

They also had another piece of information: for each patient they knew, from internal hospital data, what the average waiting time in A&E was at the time they arrived. This means that they were able to compare the odds of death for patients discharged when the average wait in A&E was less than four hours (or more) against the odds of death for patients discharged when the wait was less than one hour. Remember, this isn't the time that individual patient waited, it's the average wait in the department, as a proxy for how frantic things were.

The results were as you might fear. For patients sent home who attended an A&E department when the average wait there was more than six hours, the odds of death were almost twice those of patients sent home when the wait was less than one hour. This odds ratio was similar for patients measured as high or low urgency at triage, so it's true for patients with both serious and less serious presentations.

Even more starkly, there's a very clear trend in the data, where each step up in waiting time results in a higher risk of death. This becomes statistically significant when average waits reach just three hours. For those who care about saving money, the odds of being admitted – and so taking up an expensive hospital bed – also rose dramatically as average wait time increased.

However important you might find those specific results, the methodological issues are much more interesting, and they all arise because of the big numbers involved. We would never have discovered any of this without huge numbers of patients' records, because the outcomes involved are rare: you only see a handful of deaths out of every 10,000 people sent home from A&E.

What's more, because they had so many patients' data, the researchers were able to see an effect even within hospitals, over time: so it wasn't just that crap hospitals overall had longer waits, and higher death rates. What's more, amazingly, they didn't lose a single patient during follow-up: the death – or otherwise – of every single patient who was sent home from A&E could be tracked through their notes.

No individual patient or doctor could possibly have shown with any certainty, from their own personal experience of any one adverse outcome, that long waiting times in A&E are dangerous. This study is a remarkable testament to the power of good-quality computerised health records, and the kinds of new knowledge you can generate from interrogating them. It's also, I'll agree, a pretty frightening result.

Give Us the Data

Guardian, 7 October 2011

Bad things happen when problems are protected by a force field of tediousness. Here is an example. Data is the fabric of the modern world: just as we walk down pavements, so we trace routes through data, and build knowledge and products out of it. The government has lots of data that has already been collected, because it was needed to run the country properly: simple stuff like maps, postcode areas, land ownership, procurement data, endless weather readings, and so on.

Right now a fight is happening in Whitehall between two factions in government: one group thinks we should give this data away for free, as a matter of principle, because it will make good things happen; the other thinks we should restrict access, and sell it. A consultation is under way. Despite a positive ministerial introduction, each of the three options it gives for releasing data is foolishly restrictive. Here's why that's a problem.

As things stand, much everyday government data is locked down so hard that nerds are forbidden to repurpose it. You could have a map of who owns what in your town, on your screen, at a click. You could find out what company boards someone sits on, and map their relationships and overlaps with all the other directors in the country. You could download transcripts of court proceedings that affect you. All this is blocked by the government's restrictive data policies.

There are areas where access has been won by the shame of a simple moral argument. Hansard is a record of everything that happens in Parliament. TheyWorkForYou.com is a repurposing of that data which adds huge value, not just by being more usable than Hansard, but by identifying patterns in MPs' voting behaviour. When it first came out, Hansard argued – embarrassingly – that this was an illegal breach of copyright.

But there are also straight commercial applications. If you're making services or things that you sell to government, then seeing what they use and need helps you sell them stuff. That data is even internally useful: if you can see what everyone else is paying for toilet paper, you might get a better deal for your own department.

All this data has to be created, regardless of whether or not it gets sold, simply in order to run the country. You could 'sweat the asset', and charge money for access; but if you release it for free, at barely any cost to yourself, without fiddliness, in its raw form, the benefits are potentially huge.

This becomes especially clear when you notice how the restrictions extend beyond specific realms of data, and into the kind of core structural information that is needed as a civic skeleton for simple, everyday activity. The Royal Mail still owns all our postcode information, and you can't get the house-number boundaries of each specific postcode without paying. All the most interesting data projects involve linking one dataset with another, and for addresses, that often means using post-codes, as a commonly used structural spine (I'm willing to bet that you don't know your house's latitude and longitude). This kind of framework data is the pavement of data space, and if you're not allowed to use it, projects go unmade.

The economic loss is almost impossible to measure: if any of the projects I've already described sound trivial to you, remember that this is a crippled field, where innovators have barely had a chance to get their eyes in. Amazing things happen when you pull individual pieces of information together into larger linked datasets: meaning emerges, as you produce facts from figures. If you've ever wished you were born in the nineteenth century, when there were so many obvious inventions and ideas to hook for yourself, then I seriously recommend you become a coder, because future nerds will look back on this time with the exact same envy. But that leap forward will be tediously retarded if we don't make the government allow us to use the pavements.

Care.data Can Save Lives:
But Not If We Bungle It

Guardian, 21 February 2014

Everything would be much simpler if science really was 'just another kind of religion'. But medical knowledge doesn't appear out of nowhere, and there is no ancient text to guide us. Instead, we learn how to save lives by studying huge datasets on the medical histories of millions of people. This information helps us identify the causes of cancer and heart disease; it helps us spot side effects from beneficial treatments, and switch patients to the safest drugs; it helps us spot failing hospitals, or rubbish surgeons; and it helps us spot the areas of greatest need in the NHS. Numbers in medicine are not an abstract academic game: they are made of flesh and blood, and they show us how to prevent unnecessary pain, suffering and death.

All this vital work is now being put at risk by the bungled implementation of the care.data project. It was supposed to link all NHS data about all patients together into one giant database, like the one we already have for hospital episodes; instead it has been put on hold for six months, in the face of plummeting public support. It should have been a breeze. But we have seen arrogant paternalism, crass boasts about commercial profits, a lack of clear governance, and a failure to communicate basic science properly. All this has left the field open for wild conspiracy theories. It would take very little to fix this mess, but time is short, and lives are at stake.

The care.data project was promoted in two ways: we will use your data for lifesaving research; and we will give it to the private sector for commercial exploitation, creating billions for the UK economy. This marriage was a clear mistake: by and large, the public support public research, but are nervous about commercial exploitation of their health data.

Now the teams behind care.data are trying to row back, explaining that access will only be granted for research that benefits NHS patients. That is laudable, but it is potentially a very broad notion. It's one we would want to unpack, with clear, worked examples of the kind of things that would be permitted, and the kind of things that would be refused. But that's not possible because, bizarrely, the specific principles, guidelines, committees and regulations that will determine all these decisions have not yet been clearly set out. This poses several difficulties. Firstly, the public are being asked to support something that feels intuitively scary, about the privacy of their medical records, without being told the details of how it will work. Secondly, the field has been left open to conspiracy theories, which are hard to refute without concrete guidance on how permissions for access really will work.

That said, many criticisms have been absurd. There has been endless discussion around the idea of health insurers buying health records, for example, and using them to reject high-risk patients. Call an insurer right now and see how you get on: within minutes you will be asked to declare your full medical history, waive confidentiality and grant access to your full medical notes anyway.

Many have complained about drug companies getting access to data, and this is more complex. On the one hand, arrangements like these are long-standing and essential: if medicines regulators get a few unusual side-effect reports from patients, they go to the drug company and force them to do a big study, examining – for example – 10,000 patients' records, to find out if people on that drug really do have more heart attacks than we'd expect. To do this, the UK health regulator itself sells industry the data, in the past from something called the GP Research Database, which holds millions of people's records already. This needs to happen, and it's good.

But equally, people know – I've certainly shouted about it for long enough – that pharmaceutical companies also misuse data: they hide the results of clinical trials when it suits them, quite legally; they monitor individual doctors' prescribing patterns to guide their marketing efforts, and so on. The public don't trust the pharmaceutical industry unconditionally, and they're right not to.

Trust, of course, is key here, and that's currently in short supply. The NSA leaks showed us that governments were casually helping themselves to our private data. They also showed us that leaks are hard to control, because the National Security Agency of the wealthiest country in the world was unable to stop one young contractor stealing thousands of its most highly sensitive and embarrassing documents.

But there is a more specific reason why it is hard to give the team behind care.data our blind faith: they have been caught red-handed giving false reassurance on the very real – albeit modest – privacy threats posed by the system.

Tim Kelsey is the man running the show: an ex-journalist, passionate and engaging, he has drunk more open-data Kool-Aid than anyone I've ever met. He has evangelised the commercial benefits of sharing NHS data – perhaps because he made millions from setting up a hospital-ranking website with Dr Foster Intelligence – but he is also admirably evangelical about the power of data and transparency to spot problems and drive up standards. Unfortunately, he gets carried away, stepping up and announcing boldly that no identifiable patient data will leave the Health and Social Care Information Centre. Others supporting the scheme have done the same.

This is false reassurance, and that is poison in medicine, or in any field where you are trying to earn public trust. The data will be 'pseudonymised' before it is released to any applicant company, with postcodes, names and birthdays removed. But re-identifying you from that data is more than possible. Here's one example: I had twins last year (it's great; it's also partly why I've been writing less). There are 12,000 dads with similar luck each year; let's say 2,000 in London; let's say one hundred of those are aged thirty-nine, like me. From my brief online bio you can work out that I moved from Oxford to London in about 1995. Congratulations: you've now uniquely identified my health record, without using my name, postcode, or anything 'identifiable'. Now you've found the rows of data that describe my contacts with health services, you can also find out if I have any medical problems that some might consider embarrassing: incontinence, perhaps, or mental health difficulties. Then you can use that information to try to smear me: a routine occurrence if you do the work I do, whether it's done by big drug companies or dreary little quacks.

This risk isn't necessarily big, but to say it doesn't exist is crass: it's false reassurance, which ultimately undermines trust. It's also unnecessary and counterproductive, like hiding information on side effects instead of discussing them proportionately. To the best of my knowledge, we've never yet had a serious data leak from a medical research database, and there are plenty around already; but then, we are standing on the verge of a significant increase in the number of people accessing and using medical data. There are steps we can take to minimise the risks: only release a subset of the 60 million UK population to each applicant; only give out the smallest possible amount of information on each patient whose records you are sharing; suggest that people come to your data centre to run their analyses, instead of downloading records, and so on. But, while the care.data project might be planning to do some of those things, the ground rules haven't been properly written out yet.

In any case, even safeguards such as these can be worked around. There are companies out there operating in the grey areas of the law, aggregating data from every source and leak they can find, generating huge, linked datasets with information from direct-marketing lists, online purchases, mobile-phone companies and more. Who's to know if someone will start quietly aggregating all the small chunks of our health data?

This, of course, would be illegal. As Tim Kelsey and others are keen to point out, re-identifying or leaking data in any way would be a 'criminal offence'. But as this project lands, we're all becoming rapidly aware that incompetence, malice and creepiness around confidential data are policed with a worryingly light touch. Private investigators have little trouble obtaining confidential data from staff in the police force, banks and tax offices, for example.

Here's why: it took a long time for anyone to realise that Steve Tennison, a finance manager in a GP practice, had accessed patients' records on 2,023 occasions over the course of a year, although it was relevant to his work on only three occasions. The majority of the records he snooped on belonged to young women: he repeatedly accessed the records of one woman he had gone to school with, and also the records of her son. The maximum penalty for this is a fine, with a ceiling of £5,000 in magistrates' courts. Tennison was fined £996 in December 2013. This is why the public feel nervous, and this is what we need to fix.

It's painful for me to write critically about a project like care. data, because I love medical data, and I know the good it can do. We have a golden opportunity in the UK, with 60 million people cared for in one glorious NHS. Opt-outs would destroy the data, and the growing calls for an opt-in system would be worse: opt-in killed people by holding back organ donation, and more than that, it would exacerbate social inequality around data, because the poorest patients, who are the most likely to be unwell, are also the least engaged with services, the least likely to opt in. They would become invisible.

So here's my advice: if you're thinking of opting out – wait. If you run care.data – listen. There are three things the government can do to rescue this project.

Firstly, make a proper announcement about what you will do during the six-month delay. You cannot rely on blind trust when it comes to sharing private medical records, so explain that you'll be coming back soon with a clear story. Sort out the governance framework, present unambiguous rules and principles explaining how data will be shared, list the specific clinical codes you're proposing to upload, then give real-world examples of the kind of access applications that would be approved, and the kind that would be rejected. This is fair, and sensible.

Secondly, show the public how lives are saved by medical research. This needs examples from the vast archives of medical research on cancer, heart disease and more. Alongside that, give a clear nod to the small risks, and an explanation of how they will be mitigated. Never be seen to give false reassurance on these risks; if you do, you will lose patients' trust forever.

Lastly, we need stiff penalties for infringing medical privacy, on a grand and sadistic scale. Fines, like parking tickets, are useless for individuals and companies: anyone leaking or misusing personal medical data needs a prison sentence, as does their CEO. Their company – and all its subsidiaries – should be banned from accessing medical data for a decade. Rush some test cases through, and hang the bodies in the town square.

If the government does all this, it has a good chance of saving a vital data project, and permitting medical research that saves vast numbers of lives to continue. If the government tries to fudge – with half measures, superficial PR and false reassurance – then care.data will fail, and it might well bring down other sensible public health research with it. Lives are at stake. This cannot be left to the last minute in the six-month pause, and time is precious. It's February. If you're thinking of opting out, please don't. But mark your diary for May.

Care.data Has Been Bungled

Guardian, 28 February 2014

I am embarrassed. Last week I wrote in support of the government's plans to collect and share the medical records of all patients in the NHS, albeit with massive caveats. The research opportunities are huge, but we already knew that the implementation was chaotic, with poor public information, partly because the checks and balances on who gets access to data – and how – have not yet been devised or implemented. When you're proposing to share our most private medical records, vague promises and an imaginary regulatory framework are not reassuring.

Now it's worse. On Monday, the Health and Social Care Information Centre admitted giving the insurance industry the coded hospital records of millions of patients, pseudonymised, but re-identifiable by anyone with malicious intent, as I explained last week. These were crunched by actuaries into tables showing the likelihood of death, depending on various features such as age or disease, to help inform insurance premiums.

We can reasonably disagree on whether you find this use of your medical records acceptable, but the process must be competent and transparent. The HSCIC has now told the BBC that this release of your medical records broke the rules, and that there may have been other similarly erroneous releases; but it won't say more until 'later this year'.

On Tuesday, at a Health Select Committee hearing, things got worse. The HSCIC said it couldn't share documentation on these releases because it had all been done by its predecessor body, the NHS Information Centre – even though the HSCIC replaced the NHSIC in 2013, and is in the same building, doing the same job, with almost identical personnel and all the old records. Furthermore, the actuaries' report using the hospital data carries the HSCIC's logo – not the old NHSIC one – with the HSCIC's admitted full consent. If HSCIC disapproves of NHSIC releasing this data – or regards it as illegal – why did it add its logo and approval to the output?

Also, is it really true that release to the insurance industry is unacceptable to the HSCIC? Its own information governance assessment from August says that access to individual patients' records can 'enable insurance companies to accurately calculate actuarial risk so as to offer fair premiums to its [sic] customers. Such outcomes are an important aim of Open Data, an important government policy initiative.' Is that document binding? What are the rules? Are there previous dodgy data-sharing arrangements, agreed by the NHSIC, that the HSCIC is still honouring, with data still flowing out of the building?

This is chaos. Then, on Thursday, to make things worse, Public Health Minister Jane Ellison appears to have misled Parliament, telling it that the data released by the HSCIC was 'publicly available, non-identifiable and in aggregate form'. This is utterly untrue. It was line-by-line data – every individual hospital episode, for every individual patient, with unique pseudonymous identifiers – which was then aggregated into summary tables by the actuaries.

To summarise, a government body handed over parts of my medical records to people I've never met, outside the NHS and the medical research community, but it is refusing to tell me what it handed over, or who it gave it to, and the Minister is now incorrectly claiming that it never happened anyway.

There are people in my profession who think they can ignore this problem. Some are murmuring that this mess is like MMR, a public misunderstanding to be corrected with better PR. They are wrong: it's like nuclear power. Medical data, rarefied and condensed, presents huge power to do good, but it also presents huge risks. When leaked, it cannot be unleaked; when lost, public trust will take decades to regain.

This breaks my heart. I love big medical datasets, I work on them in my day job, and I can think of a hundred life-saving uses for better ones. But patients' medical records contain secrets, and we owe them our highest protection. Where we use them – and we have used them, as researchers, for decades without a leak – this must be done safely, accountably and transparently. New primary legislation, governing who has access to what, must be written: but that's not enough. We also need vicious penalties for anyone leaking medical records; and the HSCIC needs to regain trust, by releasing all documentation on all past releases, urgently. Care.data needs to work: in medicine, data saves lives.

The care.data programme was suspended shortly after this piece was published, with the promise that they'd have a think and relaunch in six months. Six months have already passed, and there has been no relaunch. I'm on their Advisory Group and continue to shout about the issues raised above, indoors and out. Medical data can save lives, but if the single biggest project ever conceived on patient records is not handled properly, we risk destroying public trust for all such projects, not just care.data.

SURVEYS

The Huff

Guardian, 19 January 2008

In 1954 a man called Darrell Huff published a book called *How to Lie with Statistics*. Chapter 1 is called 'The Sample With Built-In Bias', and it reads exactly like this column, which I'm about to write, on a *Daily Telegraph* story in 2008.

Huff sets up his headline: 'The Average Yaleman, Class of 1924, Makes $25,111 a Year!' said *Time* magazine, half a century ago. That figure sounded pretty high: Huff chases it, and points out the flaws. How did they find all these people they asked? Who did they miss? Losers tend to drop off the alma mater radar, whereas successful people are in *Who's Who* and the *College Record*. Did this introduce 'selection bias' into the sample? And how did they pose the question? Can that really be salary rather than investment income? Can you trust people when they self-declare their income? Is the figure spuriously precise? And so on.

In the intervening fifty years this book has sold one and a half million copies. It's the greatest-selling stats book of all time (tough market), and it remains in print, at just £8.99.

Meanwhile, 'Doctors say no to abortions in their surgeries' is the headline in the *Daily Telegraph*. 'Family doctors are threatening a revolt against Government plans to allow them to perform abortions in their surgeries, the *Daily Telegraph* can disclose.' A revolt? 'Four out of five GPs do not want to carry out terminations even though the idea is being tested in NHS pilot schemes, a survey has revealed.'

Channelling Huff through my fingers, in a trancelike state, I went in search of the figures. Was this a systematic survey of all GPs, with lots of chasing to catch the non-responders? Telephoning them at work? A postal survey, at least? No. It was an informal poll through doctors.net.uk, an online chat site for doctors, producing this major news story about a profession threatening a revolt.

The statement to which doctors were invited to respond was this: 'GPs should carry out abortions in their surgeries'. You can 'strongly agree', 'agree', 'don't know', 'disagree' or 'strongly disagree'.

I might be slow, but I myself do not fully understand the statement. Is that 'should' as in 'should', as in, 'ought to', as in 'coerced'? And in what circumstances? With extra training, time, and money? With extra systems in place for adverse outcomes? This is a chat website where doctors go to grumble, cynically, in good company. Are they saying 'no' because this new responsibility would involve more work and lower morale? Would you even click the 'abortion' link in the chat pages index if you didn't already have an interest in abortion?

And stepping bravely beyond the second word 'should', what does 'carry out abortions in their surgeries' mean? Looking at the comments in the chat forum – as I am doing right now – plenty of the doctors seemed to think the question referred to surgical abortions, not the relatively safe oral pill for termination of early pregnancy. Doctors aren't all that bright, you see, and questionnaire respondents in general may not necessarily know what you're thinking about if you don't write a proper question.

Here are some quotes from the doctors in the discussion underneath this poll. 'This is a preposterous idea. How can GPs ever carry out abortions in their own surgeries. What if there was a major complication like uterine and bowel perforation?' 'The only way it would or rather should happen is if GP practices have a surgical day care facility as part of their premises which is staffed by appropriately trained staff, i.e. theatre staff, anaesthetist and gynaecologist . . . any surgical operation is not without its risks, and presumably [we] will undergo gynaecological surgical training in order to perform.' 'What are we all going on about? Let's all carry out abortions in our surgeries, living rooms, kitchens, garages, corner shops, you know, just like in the old days.'

But my favourite is this: 'I think that the question is poorly worded and I hope that DNUK do not release the results of this poll to the *Daily Telegraph*.'

A New and Interesting
Form of Wrong

Guardian, 27 November 2010

Wrong isn't enough: we need interestingly wrong, and this week that came in some research from Stonewall, an organisation for which I generally have great respect, which was reported in the *Guardian*. Stonewall has conducted a survey, and its press release says it shows that 'the average coming-out age has fallen by over twenty years'.

People may well be coming out earlier than before – intuitively, that seems plausible – but Stonewall's survey is flawed by design, and contains some interesting statistical traps.

Through social networking sites, Stonewall asked 1,536 people – who were already out – how old they were when they came out. Among the over-sixties, the average age was thirty-seven; those in their thirties had come out at an average age of twenty-one; in the group aged eighteen to twenty-four, the average age for coming out was seventeen.

Why is the age coming down? Here's one reason. Obviously, there are no out gay people in the eighteen-to-twenty-four group who came out at an age later than twenty-four; so the average age at which people in the eighteen-to-twenty-four group came out cannot possibly be greater than the average age of that group, and certainly it will be lower than, say, thirty-seven, the average age at which people in their sixties came out.

For the same reason, it's very likely indeed that the average age of coming out will increase as the average age of each age group rises. In fact, if we assume (in formal terms we could call this a 'statistical model') that at any time, all the people who are out have always come out at a uniform rate between the age of ten and their current age, you would get almost exactly the same figures (you'd get fifteen, twenty-three and thirty-five, instead of seventeen, twenty-one and thirty-seven). This is almost certainly why 'the average coming-out age has fallen by over twenty years': in fact you could say that Stonewall's survey has found that on average, as people get older, they get older.

But there is also an interesting problem around whether, with the data it collected, Stonewall could ever have created a meaningful answer to the question 'Have people started coming out earlier?' It's a difficult analysis to design, because in each age band there is no information on gay people who are not yet out, but may come out later, and also it's hard to compare each age band with the others.

You could try to fix this by restricting all the data to include only those people who came out under the age of twenty-four, and then measure the mean age of coming out for each age group (eighteen-to-twenty-four, thirties, sixty plus) in this subgroup alone. That would give you some kind of answer for this very narrow age band, but even that makes some very dubious statistical assumptions. And if we allowed ourselves this move, we'd then be working with an extremely small set of data: only thirty-three respondents aged over sixty, for example.

Even then, the discussion of this poll also assumes that the age at which people know their sexuality has remained unchanged. Some believe that everyone's sexuality is fixed and known from birth – I may be walking into a minefield here – but if the age at which people recognise their own sexuality is changing, then a more relevant figure by which to measure discomfort at coming out might be the delay, rather than the absolute age.

I thought I'd already covered all the ways that a survey could get things wrong, but this one brought something new. Maybe we should accept that all research of this kind is only produced as a hook for a news story about a political issue, and isn't ever supposed to be taken seriously. In any case, my intuition is that a well-constructed study would probably confirm Stonewall's original hypothesis. But it's still fun to dig.

'Hello Madam, Would You Like Your Children to Be Unemployed?'

Guardian, 20 November 2010

Obviously I like nerdy days out: like Kelvedon Hatch secret nuclear bunker, maybe, with its sign on the A128 pointing the way to the 'Secret Nuclear Bunker'. Last month eight of us commissioned a boat to get onto a rotting man-made World War II sea-fort in the middle of the ocean through Project Redsand (we genuinely thought we might die climbing the

ladders), and a couple of weeks earlier, myself and Mrs Bad
Science travelled to Dungeness, where a toytown narrow-gauge
railway takes you through amusement parks and back gardens,
past Derek Jarman's house, then into barren wasteland, before
depositing you incongruously at the base of a magnificent,
enormous, and terrifying nuclear power station.*

I tell you this, because I should declare an interest: I quite
like nuclear power stations, not just because they're clever, or
even because I regretfully concede they might be one of our
least bad options for power. I secretly like nuclear power stations
because they remind me, in the way nostalgia makes us pine for
things we disliked at the time, of a childhood in the early 1980s
when I knew that I would definitely die in a nuclear holocaust.

So. Last month energy company EDF conducted a poll on
whether people near Hinkley Point nuclear power station would
like it to be expanded. The BBC dutifully reported the results:

---

* See page 379.

'EDF Survey Shows Support for Hinkley Power Station', said the headline. 'Six in 10 people support a new power station at Hinkley'. Polls like this convince locals, and politicians.

But Leo Barasi at the blog ClimateSock has obtained the original polling questions from ICM, and found a masterclass in manipulation.

First, respondents are set into the frame with a simple starter: 'How favourable or unfavourable is your opinion of the nuclear energy industry?' Then things heat up. 'To what extent do you agree or disagree with the following statement: Nuclear energy has disadvantages but the country needs nuclear power as part of the energy balance with coal, gas and wind power.' As Leo says, this is structured in a way that makes it harder to disagree. 'It appears reasoned: taking on board the downsides of nuclear before drawing a measured conclusion that it's a necessary evil to produce a greater good.' As a result, only 13 per cent disagree, but the whole audience is gently nudged.

Then locals are asked a whole series of branching questions, forcing them to weigh up the positive and negative impacts a new power station would have on the area. People who think it would be positive are asked to also weigh up the negative, and people who think it would be negative are asked to weigh up the positive factors, and everyone is asked to say why they think what they think.

Then, in a killer move, they're asked: 'How important, if at all, do you consider a new power station at Hinkley to each of the following? To the creation of local jobs? To the future of local businesses?' And take a moment to reinforce those concerns: 'Why do you say that?'

Finally, after being led on this thoughtful journey, and immediately after mulling over the beneficial economic impact it would have on their community, the locals are asked if they're in favour of a new nuclear power station. It's the results of this, the final question, that are reported in the press release and headlines.

To me it seems clear that this long series of preceding questions will guide people down a very specific path when thinking about a nuclear power station. It's a guided narrative, and that might make sense if you were trying to advocate a kind of structured decision-making, but it's very unlikely to produce results that reflect the true range of local views, partly because we're all a bit thoughtless in the real world, and follow our guts in odd ways; but partly because the penultimate question is 'Do you want your children to be unemployed?' rather than 'Are you secretly terrified we might cock up and give you cancer?'

So I still quite like nuclear power stations, but more than that, as ever, I salute the PR industry for finding new and elaborate ways to muddy the waters. And I salute the nerds who bust them for it.

EPIDEMIOLOGY

Beau Funnel

Guardian, 28 October 2011

The BBC has found a story: '"Threefold variation" in UK bowel cancer rates'. The average death rate across the UK from bowel cancer is 17.9 per 100,000 people, but in some places it's as low as nine, and in some places it's as high as thirty. What can be causing this?

Journalists tend to find imaginary patterns in statistical noise, which we've covered many times before. But this case is particularly silly, as you will see, and it has a heartwarming, nerdy twist.

Paul Barden is a quantitative analyst. He saw the story, and decided to download the data and analyse it himself. The claims come from a press release by the charity Beating Bowel Cancer: they've built a map where you can find your own local authority's bowel cancer mortality rate and get worried, or reassured. Using a 'scraping' program, Barden brought up the page for each area in turn, and downloaded the figures. By doing this he could make a spreadsheet showing the death rate in each region, and its population. From here things get slightly complicated, but very rewarding.

We know that there will be random variation around the average mortality rate, and also that this will be different in different regions: local authorities with larger populations will have less random variation than areas with smaller populations, because the variation from chance events gets evened out more when there are more people.

You can show this formally. The random variation for this kind of mortality rate will follow the Poisson distribution (a bit like the bell-shaped curve you'll be familiar with). This bell-shaped curve gets narrower – less random variation – for areas with a large population.

So, Barden ran a series of simulations in Excel, where he took the UK average bowel cancer mortality rate and a series of typical population sizes, and then used the Poisson distribution to generate figures for the bowel cancer death rate that varied with the randomness you would expect from chance.

This random variation predicted by the Poisson distribution – before you even look at the real variations between areas – shows that you would expect some areas to have a death rate of seven, and some areas to have a death rate of thirty-two. So it turns out that the real UK variation, from nine to thirty-one, may actually be *less* than you'd expect from chance.

Then Barden sent his blog to David Spiegelhalter, a Professor of Statistics at Cambridge, who runs the excellent website Understanding Uncertainty. Spiegelhalter suggested that Barden could present the real cancer figures as a funnel plot, and that's what you see opposite.

I cannot begin to tell you how happy it makes me that Spiegelhalter, author of *Funnel Plots for Comparing Institutional Performance* – the citation classic from 2005 – can be found by a random blogger online, and then collaborate to make an informative graph of some data that's been over-interpreted by the BBC.

But back to the picture. Each dot is a local authority. The dots higher up show areas with more deaths. The dots further to the right show ones with larger populations. As you can see,

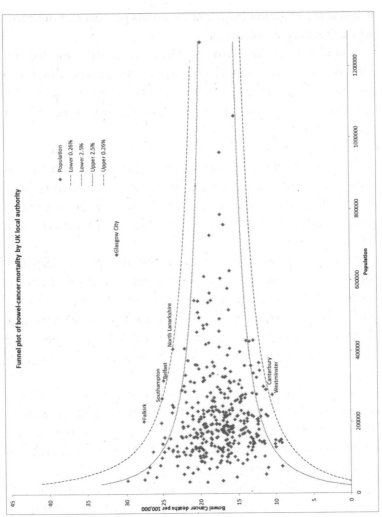

areas with larger populations are more tightly clustered around the UK average death rate, because there's less random variation in bigger populations. Lastly, the dotted lines show you the amount of random variation you would expect to see, from the Poisson distribution, and there are very few outliers (well, one main one, really).

Excitingly, you can also do this yourself online. The Public Health Observatories provide several neat tools for analysing data, and one will draw a funnel plot for you, from exactly this kind of mortality data. The bowel cancer numbers are in the table above. You can paste them into the Observatories' tool, click 'calculate', and experience the thrill of touching real data.

In fact, if you're a journalist, and you find yourself wanting to claim one region is worse than another, for any similar set of death rate figures, then do feel free to use this tool on those figures yourself. It might take five minutes.

The week after this column came out, a letter was published from Gary Smith, UK News Editor at the BBC. It said: 'The BBC stands by this report as an accurate representation of the figures, which were provided by the reputable charity Beating Bowel Cancer. Dr Goldacre suggests the difference between the best- and worst-performing authorities falls within a range that could be expected through change [I don't suggest this: I demonstrate it to be true]. But that does not change the fact that there is a threefold difference between the best and worst local authorities.' This is a good example of the Kruger-Dunning effect: the phenomenon of being too stupid to know how stupid you're being (discussed, if you're keen to know more, in Bad Science, *page 284).*

When Journalists Do Primary Research

Guardian, 9 April 2011

This week some journalists found a pattern in some data, and ascribed a cause to it. 'Recession linked to huge rise in antidepressants' said the *Telegraph*. 'Economic woes fuel dramatic rise in use of antidepressants' said the *Daily Mail*. 'Record numbers of people are being handed antidepressants' said the *Express*. Even the *Guardian* joined in. It seems to have come from a BBC report.

The journalists are keen for you to know that these figures come from a Freedom of Information Act request, which surprised me, since each year – like you – I enjoy reading the Prescription Cost Analysis documents, which detail everything that has been prescribed over the previous year. The 2009 data was published in April 2010, so I guess the 2010 data was due about now.

But are the numbers correct? Yes. From 2006 to 2010 there was a 43 per cent increase in the number of prescriptions for the SSRI class of antidepressants. Does that mean more people are depressed in the recession?

Firstly, this rise in scripts for antidepressants isn't a new phenomenon. In 2009 the *BMJ* published a paper titled 'Explaining the rise in antidepressant prescribing', which looked at the period from 1993 to 2005. In the five years from 2000 to 2005 – the boom before the bust these journalists are writing about – antidepressant prescribing also increased, by 36 per cent. That isn't very different from 43 per cent, so it feels unlikely that the present increase in prescriptions is due to the recession.

That's not the only problem here. It turns out that the number of prescriptions for an SSRI drug is a pretty unhelpful way of measuring how many people are being treated for depression: not just because people get prescribed SSRIs for all kinds of other things, like anxiety, PTSD, hot flushes, and more; and not just because doctors have moved away from older types of antidepressants, so they would be prescribing more of the newer SSRI drugs even if the number of people with depression had stayed the same.

Excitingly, it's a bit more complicated than that. A 2006 paper from the *British Journal of General Practice* looked at prescribing and diagnosis rates in Scotland. Overall, again, the number of prescriptions for antidepressants increased, from 1.5 million in 1996 to 2.8 million in 2001 (that is, it almost doubled).

But the researchers of this paper also found a mystery: looking at the Scottish Health Survey, they found no increase in the prevalence of depression; and looking at the GP consultations dataset, they again found no evidence that people were presenting more frequently to their GP with depression, or that GPs were making more diagnoses of depression.

So why were antidepressant prescriptions going up? This puzzle received some kind of explanation in 2009. The *BMJ* paper above found the same increase in the number of prescriptions that the journalists reported this week, as I said. But they had access to more data: their analysis didn't just look at the total number of prescriptions in the country, or even the total number of people diagnosed with depression: it also looked at the prescription records of individual patients, in a dataset of over three million patients' electronic health records (with 200,000 people who experienced a first diagnosis of depression during this period).

They found that the rise in the overall number of antidepressant prescriptions was not due to increasing numbers of patients receiving antidepressants. It was almost entirely caused by one thing: a small increase in the small proportion of those patients who received treatment for longer periods of time. Numerically, people receiving treatment for long periods make up the biggest chunk of all the prescriptions written, so this small shift bumped up the overall numbers hugely.

I don't know for certain if that phenomenon explains the increase in prescriptions from 2006 to 2010, as it does for the period 2000 to 2005 (although, in the absence of work examining that question, since the increase in scripts was so similar, it does seem fairly likely). And I'm not expecting journalists to go to academic research databases to conduct large, complex, descriptive studies.

But if they are going to engage in primary research, and make dramatic causal claims – as they have done in this story – to the nation, then they could also, usefully, read through the proper work that's already been done, and consider alternative explanations for the numbers they've found.

Confound You!

Guardian, 5 August 2011

Fox News was excited: 'Unplanned children develop more slowly, study finds'. The *Telegraph* was equally shrill in its headline: 'IVF children have bigger vocabulary than unplanned children'. And the *British Medical Journal* press release drove it all: 'Children born after an unwanted pregnancy are slower to develop'.

The last two, at least, made a good effort to explain that this effect disappeared when the researchers accounted for social and demographic factors. But was there ever any point in reporting the raw finding, from before this correction was made?

I will now demonstrate, with a nerdy table illustration, how you correct for things such as social and demographic factors. You'll have to pay attention, because this is a tricky concept; but at the end, when the mystery is gone, you will see why reporting the unadjusted figures as the finding, especially in a headline, is simply wrong.

Correcting for an extra factor is best understood by doing something called 'stratification'. Imagine you do a study, and you find that people who drink are three times more likely to get lung cancer than people who don't. The results are in Table 1. Your odds of getting lung cancer as a drinker are 0.16 (that's 366 ÷ 2,300). Your odds as a non-drinker are 0.05. So your odds of getting lung cancer are three times higher as a drinker ($0.16 \div .05$ is roughly 3, and that figure is called the 'odds ratio') – as in Table 1 below.

Table 1: Everyone

	Lung cancer	No cancer	Odds of cancer
Drinker	366	2300	0.16
Non-drinker	98	1856	0.05

But then some clever person comes along and says: Wait, maybe this whole finding is confounded by the fact that drinkers are more likely to smoke cigarettes. That could be an alternative explanation for the apparent relationship between drinking and lung cancer. So you want to factor smoking out.

The way to do this is to chop your data in half, and analyse non-smokers and smokers separately. So you take only the people who smoke, and compare drinkers against non-drinkers; then you take only the people who don't smoke, and compare drinkers against non-drinkers in that group separately. You can see the results of this in the second and third tables.

Table 2: Only smokers

	Lung cancer	No cancer	Odds of cancer
Drinker	330	1100	0.30
Non-drinker	47	156	0.30
			Odds ratio = 1.0

Table 3: Only non-smokers

	Lung cancer	No cancer	Odds of cancer
Drinker	36	1200	0.03
Non-drinker	51	1700	0.03
			Odds ratio = 1.0

Now your findings are a bit weird. Suddenly, since you've split the data up by whether people are smokers or not, drinkers and non-drinkers have exactly the same odds of getting lung cancer. The apparent effect of drinking has been eradicated, and this means that the observed risk of drinking was entirely due to smoking: smokers had a higher chance of lung cancer – in fact their odds were 0.3 rather than 0.03, ten times higher – and drinkers were more likely to also be smokers. Looking at the figures in these tables, 203 out of 1,954 non-drinkers smoked, whereas 1,430 out of 2,666 drinkers smoked.

I explained all this with a theoretical example, where the odds of cancer apparently trebled before correction for smoking. Why didn't I just use the data from the unplanned pregnancies paper? Because in the real world of research, you're often correcting for lots of things at once. In the case of this *BMJ* paper, the researchers corrected for parents' socioeconomic position and qualifications, sex of child, age, language spoken at home, and a huge list of other factors.

When you're correcting for so many things, you can't use old-fashioned stratification, as I did in this simple example, because you'd be dividing your data up among so many smaller tables that some would have no people in them at all. That's why you calculate your adjusted figures using cleverer methods, such as logistic regression* and likelihood theory. But it all comes down to the same thing. In our example above, alcohol wasn't really associated with lung cancer. And in this *BMJ* paper, unplanned pregnancy wasn't really associated with slower development. Pretending otherwise is just silly.

---

* In case it's been puzzling you, epidemiologists use 'odds' (e.g. 366 ÷ 2300 for the top row of Table 1) rather than 'proportions' (which would be 366 ÷ 2666 for the top row of Table 1) because odds work more neatly when you use 'logistic regression', which is the more advanced technique mentioned above. If you're interested to know more, I recommend coming to London School of Hygiene and Tropical Medicine to do our MSc in Epidemiology.

Bicycle Helmets and the Law*

Ben Goldacre and David Spiegelhalter, *British Medical Journal*, 12 June 2013

We have both spent a large part of our working lives discussing statistics and risk with the general public. We both dread questions about bicycle helmets. The arguments are often heated and personal; but they also illustrate some of the most fascinating challenges for epidemiology, risk communication and evidence-based policy.

With regard to the use of bicycle helmets, science broadly tries to answer two main questions. At a societal level, 'What is the effect of a public health policy that requires or promotes helmets?' and at an individual level, 'What is the effect of wearing a helmet?' Both questions are methodologically challenging and contentious.

The linked paper by Dennis and colleagues (doi:10.1136/bmj.f2674) investigates the policy question and concludes that the effect of Canadian helmet legislation on hospital admission for cycling head injuries 'seems to have been minimal'. Other ecological studies have come to different conclusions, but the current study has somewhat superior methodology – controlling for background trends and modelling head injuries as a proportion of all cycling injuries.

---

* This is an editorial I wrote for the *British Medical Journal* with David Spiegelhalter about the complex, contradictory mess of evidence on the impact of bicycle helmets. Like most places where there's controversy and disagreement, this is a great opportunity to walk through the benefits and shortcomings of different epidemiological techniques, from case-control studies to modelling. Epidemiology is my day job – *Bad Science* and *Bad Pharma* are both, effectively, epidemiology textbooks with bad guys – and since the techniques of epidemiology are at the core of most media stories and squabbles on health, it's very weird that you don't hear the word more often.

This finding of 'no benefit' is superficially hard to reconcile with case-control studies, many of which have shown that people wearing helmets are less likely to have a head injury. Such findings suggest that, for individuals, helmets confer a benefit. These studies, however, are vulnerable to many methodological shortcomings. If the controls are cyclists presenting with other injuries in the emergency department, then analyses are conditional on having an accident and therefore assume that wearing a helmet does not change the overall accident risk. There are also confounding variables that are generally unmeasured and perhaps even unmeasurable. People who choose to wear bicycle helmets will probably be different from those who ride without a helmet: they may be more cautious, for example, and so less likely to have a serious head injury, regardless of their helmets.

People who are forced by legislation to wear a bicycle helmet, meanwhile, may be different again. Firstly, they may not wear the helmet correctly, seeking only to comply with the law and avoid a fine. Secondly, their behaviour may change as a consequence of wearing a helmet through 'risk compensation', a phenomenon that has been documented in many fields. One study – albeit with a single author and subject – suggests that drivers give larger clearance to cyclists without a helmet.

Even if helmets do have an effect on head-injury rates, it would not necessarily follow that legislation would have public health benefits overall. This is because of 'second-round' effects, such as changes in cycling rates, which may affect individual and population health. Modelling studies have generally concluded that regular cyclists live longer because the health effects of cycling far outweigh the risk of crashes. This trade-off depends crucially, however, on the absolute risk of an accident: any true reduction in the relative risk of head injury will have a greater impact where crashes are more common, such as for children.

The impact on all-cause mortality, and on head injuries, may be even further complicated if such legislation has varying effects on different groups. For example, a recent study identified two broad subpopulations of cyclist: 'one speed-happy group that cycle fast and have lots of cycle equipment including helmets, and one traditional kind of cyclist without much equipment, cycling slowly'. The study concluded that compulsory cycle-helmet legislation may selectively reduce cycling in the second group. There are even more complex second-round effects if each individual cyclist's safety is improved by increased cyclist density through 'safety in numbers', a phenomenon known as Smeed's law. Statistical models for the overall impact of helmet habits are therefore inevitably complex and based on speculative assumptions. This complexity seems at odds with the current official BMA policy, which confidently calls for compulsory helmet legislation.

Standing over all this methodological complexity is a layer of politics, culture and psychology. Supporters of helmets often tell vivid stories about someone they knew, or heard of, who was apparently saved from severe head injury by a helmet. Risks and benefits may be exaggerated or discounted depending on the emotional response to the idea of a helmet. For others, this is an explicitly political matter, where an emphasis on helmets reflects a seductively individualistic approach to risk management (or even 'victim blaming'), while the real gains lie elsewhere. It is certainly true that in many countries, such as Denmark and the Netherlands, cyclists have low injury rates, even though rates of cycling are high and almost no cyclists wear helmets. This seems to be achieved through interventions such as good infrastructure, stronger legislation to protect cyclists, and a culture of cycling as a popular, routine, non-sporty, non-risky behaviour.

In any case, the current uncertainty about any benefit from helmet wearing or promotion is unlikely to be substantially reduced by further research. Equally, we can be certain that helmets will continue to be debated, and at length. The enduring popularity of helmets as a proposed major intervention for increased road safety may therefore lie not with their direct benefits – which seem too modest to capture compared with other strategies – but more with the cultural, psychological and political aspects of popular debate around risk.

Screen Test

Guardian, 12 January 2008

So we're all going to get screened for our health problems, by some businessmen who've bought a CT scanner and put an advert in the paper maybe, or perhaps by Gordon Brown: because screening saves lives, data is good, and it's always better to do something rather than nothing.

Unfortunately, it's a tiny bit more complicated than that.

Screening is a fascinating area, mainly because of the maths of rare events, but also because of the ethics. Screening isn't harmless, as tests – inevitably – aren't perfect. You might get a false alarm, causing stress and anxiety ('the worst time in my life' said women in one survey on breast screening). Or you might have to endure more invasive medical investigations to follow up the early warning: even something as innocuous as a biopsy can sometimes result in harmful adverse events, and if you do a lot of those, unnecessarily, in a population, then you're hurting people, sometimes more than you're helping. Lastly, people might get false reassurance from a false negative result, and ignore other niggles, which can in turn delay the diagnosis of genuine problems.

Then, there are the interesting ethical issues. One of the proposed screening programmes is intended to catch abdominal aortic aneurysms earlier. An AAA is a swelling of the main blood-vessel trunk in your belly: they can rupture without much warning, and when they do, people often die fast and frighteningly. But if you know the AAA is there, and do the repair operation at your leisure before it ruptures, then survival is far better. Screening and repairing have been shown to reduce mortality by around 40 per cent, looking at the whole population, which is a good thing.

But remember, you will operate on some people – as a preventive measure, because you picked up their aneurysm on screening – who would *never* have died from their aneurysm: it would have just ticked away quietly, not rupturing. And some of the people you operate on unnecessarily (and remember, there's no crystal ball to identify these people) will die of complications on the operating table. They only died because of your screening programme. It saves lives *overall*, but Fred Bloggs – loving husband of Winona Bloggs – who would have lived, is now dead, thanks to you.

That's Vegas, you could say. But it's tricky, and the sums are often close. For example, mammogram screening for breast cancer every two years has been estimated to prevent two deaths per thousand women aged fifty to fifty-nine over ten years: that is good. But achieving this requires 5,000 screenings among those thousand women, resulting in 242 recalls, and sixty-four women having at least one biopsy. Five women will have cancer detected and treated. Again, this isn't an argument against screening, we're just walking through some example numbers.

Although, interestingly, that's not something everybody is keen to do with screening. People in healthcare can be zealots, and enthusiasts, and we can often project our own values and preferences onto everyone else. Researchers have studied the invitation letters sent out for screening programmes, along with the websites and pamphlets, and they have repeatedly been shown to be biased in favour of participation, and lacking in information.

Where figures are given, they generally use the most dramatic and uninformative way of expressing the benefits: the 'relative risk reduction' is given, the same statistical form that journalists prefer – for example, 'a 30 per cent reduction in deaths from breast cancer' – rather than a more informative figure like the 'number needed to screen' – say, 'two lives saved for every thousand women scanned'. Sometimes the leaflets even contain borderline porkies, like this one from Ontario: 'There has been a 26 per cent increase in breast cancer cases in the last ten years,' it said, in scary and misleading tones. This was roughly the level of over-diagnosis caused by screening over the preceding ten years during which the screening programme itself had been operating.

These problems with clear information raise interesting questions around informed consent, although seductive letters do increase uptake, and so save lives. It's tricky: on the one hand, you end up sounding like a redneck who doesn't trust the gub'mint, because screening programmes are often valuable. On the other hand, you want people to be allowed to make informed choices.

And the amazing thing is, in at least one large survey of five hundred people, even when presented with the harsh realities of the tests, people made what many would still think are the right decisions. Thirty-eight per cent had experienced at least one false-positive screening test; more than 40 per cent of these individuals described the experience as 'very scary', or 'the scariest time of my life'. But looking back, 98 per cent were glad they were screened. Most wanted to know about cancer, regardless of the implications. Two thirds said they would be tested for cancer even if nothing could be done. Chin up.

How Do You Know?

Guardian, 4 June 2011

Mobile phones 'possibly' cause brain cancer, according to a report this week from the IARC (International Agency for Research on Cancer), part of the WHO. This report has triggered over 3,000 news articles around the world. Like you, I'm not interested in marginal changes around small lifestyle risks for the risks themselves; but I am interested in the methodological issues they throw up.

First, transparency: science isn't about authoritative utterances from men in white coats, it's about showing your working. What does this report say? How do its authors reason around contradictory data? Nobody can answer those questions, because the report isn't available. Nobody you see writing confidently about it has read it. There is only a press release. Nobody at the IARC even replied to my emails requesting more information.

This isn't just irritating. Phones are a potential risk exposure where people can make a personal choice. People want information. It's in the news right now. The word 'possibly' informs nobody. How can we put flesh on that with the research that is already published, and what are the limits of the research?

The crudest data you could look at is the overall rate of different brain cancers. This hasn't changed much over time, despite an increase in mobile-phone use, but it's a crude measure, affected by lots of different things.

Ideally, we'd look at individuals, to see if greater mobile use is correlated with brain cancer, but that can be tricky. These tumours are rare – about ten cases in every 100,000 people each year – and that affects how you research them.

For common things, such as heart disease, you can take a few thousand people and measure factors you think are relevant – smoking, diet, some blood tests – then wait a few years until they get the disease. This is a 'prospective cohort study', but that approach is much less useful for studying rare outcomes, like brain tumours, because you won't get enough cases appearing in your study group to spot an association with your potential cause.

For rare diseases, you do a 'retrospective case-control study': gather lots of cases; get a control group of people who don't have the rare disease but are otherwise similar; then, finally, see if your cases are more or less likely to report being exposed to mobile phones.

This sounds fine, but such studies are vulnerable to the frailties of memory. If someone has a tumour on the left of their head, say, and you ask, 'Which side did you mostly use your phone on ten years ago?', they might think, God, yes, that's a good point, and unconsciously be more likely to inaccurately remember 'the left'. In one study on the relationship between mobile-phone use and brain tumours, ten people with brain cancer (but no controls) reported phone usage figures that worked out overall as more than twelve hours a day. This might reflect people misremembering the distant past.

Then there are other problems, such as time course: it's possible that mobile phones might cause brain cancer, but through exposure over thirty years, while we've only got data for ten or twenty years, because these devices haven't been in widespread use for long. If this is the case, then the future risk may be unknowable right now (although, to be fair, other exposures that are now known to cause a peak in problems after several decades, such as asbestos, do still have measurable effects only ten years after exposure). And then, of course, phones change over time: twenty years ago the devices had more powerful transmitters, for example. So we might get a false alarm, or false reassurance, by measuring the impact of irrelevant technology.

But lastly, as so often, there's the issue of a large increase in a small baseline risk. The absolute worst-case scenario, from the Interphone study, is this: it found that phone use overall was associated with fewer tumours, which is odd; but very, very high phone use was associated with a 40 per cent increase in tumours. If everyone used their phones that much – an extreme assumption – and the apparent relationship is a genuine one, then this would still only take you from ten brain tumour cases in 100,000 people to fourteen cases in 100,000 people.

That's what 'possible' looks like: the risk itself is much less interesting than the science behind it.

Anecdotes Are Great. If They Really Illustrate the Data

Guardian, 29 July 2011

On Channel 4 News, scientists have found a new treatment for Duchenne's muscular dystrophy. 'A study in the *Lancet* today shows a drug injected weekly for three months appears to have reduced the symptoms,' they say. 'While it's not a cure, it does appear to reduce the symptoms.'

Unfortunately, the study shows no such thing. The gene for making a muscle protein called dystrophin is damaged in patients with DMD. The *Lancet* paper shows that a new treatment led to some restoration of dystrophin production in some children in a small, unblinded study.

That's not the same as symptoms improving. But Channel 4 reiterates its case, with the mother of two participants in the study. 'I think for Jack . . . it maintained his mobility . . . with Tom, there's definitely significant changes . . . more energy, he's less fatigued.'

Where did these positive anecdotes come from? Disappointingly, they come from the Great Ormond Street Hospital press release (which was tracked down online by evidence-based policy wonk Evan Harris). It summarises the dystrophin results accurately, but then, once more, it presents an anecdotal case study going way further: 'Our whole family noticed a marked difference in their quality of life and mobility over that period. We feel it helped prolong Jack's mobility and Tom has been considerably less fatigued.'

There are two issues here. Firstly, anecdotes are a great communication tool, but only when they accurately illustrate the data. The anecdotes here plainly go beyond that. Great Ormond Street denies that this is problematic (though it has changed its press release online). I strongly disagree (and this is not, of course, the first time an academic press release has been suboptimal).

But this story is also a reminder that we should always be cautious with 'surrogate' outcomes. The biological change measured was important, and good grounds for optimism, because it shows the treatment is doing what it should do in the body. But things that work in theory do not always work in practice, and while a measurable biological indicator is a hint that something is working, such outcomes can often be misleading.

Examples are easy to find, and from some of the biggest diseases in medicine. The ALLHAT trial was a vast scientific project, comparing various blood-pressure and lipid-lowering drugs against each other. One part compared 9,000 patients on doxazosin against 15,000 on chlorthalidone. Both drugs were known to lower blood pressure, to pretty much the same extent, and so people assumed they would also be fairly similar in their impact on real-world outcomes that matter, like strokes and heart attacks.

But patients on doxazosin turned out to have a *higher* risk of stroke, and cardiovascular problems, than patients on chlorthalidone – even though both lowered blood pressure – to such an extent that the trial had to be stopped early. Blood pressure, in this case, was not a reliable surrogate outcome for assessing the drug's benefits on real-world outcomes.

This is not an isolated example. A blood test called HbA1c is often used to monitor progress in diabetes, because it gives an indicator of blood-glucose levels over the preceding few weeks. Many drugs, such as rosiglitazone, have been licensed on the grounds that they reduce your HbA1c level. But this, again, is just a surrogate outcome: what we really care about in diabetes are real-world outcomes like heart attacks and death. And when these were finally measured, it turned out that rosiglitazone – while lowering HbA1c levels very well – also, unfortunately, massively increased your risk of heart attack. (The drug has now been suspended from the market.)

We might all wish otherwise, but blood tests are a mixed bag. Positive improvements on surrogate biological outcomes that can be measured in a laboratory might give us strong hints on whether something works, but the proof, ultimately, is whether we can show an impact on patients' pain, suffering, disability and death. I hope this new DMD treatment does turn out to be effective: but that's not an excuse for over-claiming, and even for the most well-established surrogate measures and drugs, laboratory endpoints have often turned out to be very misleading. People writing press releases, and shepherding misleading patient anecdotes into our living rooms, might want to bear that in mind.

Six weeks after the piece above was published, the Great Ormond Street in-house magazine RoundAbout *published a response from Professor Andrew Copp, Director of the Institute of Child Health. It seems to me that he has missed the point entirely – both on surrogate outcomes, and on the issue of widely reported anecdotal claims that went way beyond the actual data – but his words are reproduced in full below, so that you can decide for yourself.*

The end of July saw the publication in *The Lancet* of an important paper by Sebahattin Cirak, Francesco Muntoni and colleagues in the Dubowitz Neuromuscular Centre at the ICH/ Great Ormond Street Hospital (GOSH). Together with collaborators, the team provides evidence that a new technique called 'exon skipping' may be used in future to treat Duchenne muscular dystrophy (DMD).

DMD is a progressive, severely disabling neuromuscular disease that affects one in every 3,500 boys and leads to premature death. The cause is an alteration in the gene for dystrophin, a vital link protein in muscle. Without dystrophin, muscles become inflamed and degenerate, leading to the handicap seen in DMD. Patients with Becker muscular dystrophy also have dystrophin mutations, but their disease is much milder because the dystrophin protein, although shorter than normal, still functions quite well. By introducing short stretches of artificial DNA, it is possible to 'skip' over the damaged DNA in DMD cells and cause a shorter but otherwise functional protein to be made.

Previous studies showed this approach to work when the artificial DNA was injected directly into patients' muscles. The *Lancet* study asked whether the therapy would also work by intravenous injection, a crucial step as it would not be feasible to inject every muscle individually in clinical practice. Of the 19 boys who took part, seven showed an increase in dystrophin protein in their muscle biopsies. Importantly, there were few adverse effects, suggesting the treatment might be tolerated long term as would be necessary in DMD.

The study has been welcomed with great optimism in many parts of the media. However, I was disappointed to see an article about the research appearing in the 'Bad Science' column of the *Guardian*, written by Ben Goldacre. While he does not criticise the research in the *Lancet* paper, Goldacre accuses Channel 4 and GOSH of misleading the public about the extent of the advance for patients. The GOSH press release included comments by a parent who felt her two boys had shown improvement in mobility and less fatigue after treatment. Although this was only two sentences within a much longer and scientifically accurate press release, the sentences were seized upon as an opportunity for bad publicity for GOSH.

Goldacre spoils his argument by claiming the improvement in dystrophin protein level to be 'theoretical'. He says: 'things that work in theory do not always work in practice'. As a scientifically trained journalist, it is sad to see him confusing theory with an advance that is manifestly practical – the missing protein actually returned in the muscles! Nevertheless, it is a reminder that there are those in the media who like to cast doubt on the work of GOSH, and we must continue to be careful when conveying what we do, in order not to be misunderstood.

The Strange Case of the Magnetic Wine

Guardian, 4 December 2003

What is it about magnets that amazes the pseudoscientists so much? The good magnetic energy of my Magneto-Tex blanket will cure my back pain; but I need a Q-Link pendant to protect me from the bad magnetism created by household devices. Reader Bill Bingham (oddly enough, the guy who used to read the Shipping Forecast) sends in news of the exciting new Wine Magnet: 'Let your wine "age" several years in only 45 minutes! Place the bottle in the Wine Magnet! The Wine Magnet then creates a strong magnetic field that goes to the heart of your wine and naturally softens the bitter taste of tannins in "young" wines.'

I was previously unaware of the magnetic properties of wine, but this explains why I tend to become aligned with the earth's magnetic field after drinking more than two bottles. The general theory on wine maturation – and it warms the cockles of my heart to know there are people out there studying this – is that it's all about the polymerisation of tannins, which could conceivably be accelerated if they were all concentrated in local pockets: although surely not in forty-five minutes.

But this exciting new technology seems to be so potent – or perhaps unpatentable – that it is being flogged by at least half a dozen different companies. Cellarnot, marketing the almost identical 'Perfect Sommelier', even has personal testimonies from 'Susan' who works for the Pentagon, 'Maggie, Editor, *Vogue*', and a science professor, who did not want to be named but who, after giving a few glasses to some friends, exclaimed, 'The experiment definitely showed that the TPS is everything that it claims to be.' He's no philosopher of science. But perhaps all of these magnetic products will turn out to be interchangeable. Maybe I can even save myself some cash, and wear my MagneForce magnetic insoles ('increases circulation; reduces foot, leg and back fatigue') to improve the wine after I've drunk it.

And most strangely of all, none of these companies seems to be boasting about having done the one simple study necessary to test their wine magnets. As always, if any of them want advice on how to do the stats on a simple double-blind randomised trial (which could, after all, be done pretty robustly in one evening with fifty people) – and if they can't find a seventeen-year-old science student to hold their hand – I am at their disposal.

What Is Science? First, Magnetise Your Wine . . .

Guardian, 3 December 2005

People often ask me (pulls pensively on pipe), 'What is science?' And I reply thusly: Science is exactly what we do in this column. We take a claim, and we pull it apart to extract a clear scientific hypothesis, like 'Homeopathy makes people better faster than placebo,' or 'The Chemsol lab correctly identifies MRSA'; then we examine the experimental evidence for that hypothesis; and lastly, if there is no evidence, we devise new experiments. Science.

Back in December 2003, as part of our Bad Science Christmas Gift series, we discovered The Perfect Sommelier, an expensive wine-conditioning device available in all good department stores. In fact there are lots of devices like this for sale, including the ubiquitous Wine Magnet: 'Let your wine "age" several years in only 45 minutes! Place the bottle in the Wine Magnet! The Wine Magnet then creates a strong magnetic field that goes to the heart of your wine and naturally softens the bitter taste of tannins in "young" wines.'

At the time, I mentioned how easy it would be to devise an experiment to test whether people could tell the difference between magnetised and untreated wine. I also noted how strange it was that none of these devices' manufacturers seemed to have bothered, since it could be done in an evening with fifty people.

Now Dr James Rubin et al. of the Mobile Phones Research Unit at King's College London, have published that very study, in the esteemed *Journal of Wine Research*. They note the dearth of experimental research (quoting, chuffingly, the Bad Science column), and go on: 'One retailer states, "We challenge you to try it yourself – you won't believe the difference it can make."'

Unwise words.

'A review of Medline, PsychInfo, Cinahl, Embase, Amed and the Web of Science using the search term "wine and magnet" suggested that, as yet, no scientists have taken up this challenge.'

Now, this study was an extremely professional operation. Before starting, they did a power calculation: this is to decide how big your sample size needs to be, to be reasonably sure you don't miss a true positive finding by not having enough subjects to detect a small difference. Since the manufacturers' claims are dramatic, this came out at only fifty subjects.

Then they recruited their subjects, using wine. This wine had been magnetised, or not, by a third party, and the experimenters were blind to which wine was which. The subjects were also unaware of whether the wine they were tasting, which cost £2.99 a bottle, was magnetised or not. They received wine A or wine B, and it was a 'crossover design' – some people got wine A first, and some people got wine B first, in case the order you got them in affected your palate and your preferences.

There was no statistically significant difference in whether people expressed a preference for the magnetised wine or the non-magnetised wine. To translate back to the language of commercial claims: people couldn't tell the difference between magnetised and non-magnetised wine. I realise that might not come as a huge surprise to you. But the real action is in the conclusions: 'Practitioners of unconventional interventions often cite cost as a reason for not carrying out rigorous assessments of the effectiveness of their products. This double-blind randomised cross-over trial cost under £70 to conduct and took one week to design, run and analyse. Its simplicity is shown by the fact that it was run by two sixteen-year-old work experience students (EA and RI).'

'Unfortunately,' they continue, 'our research leaves us no nearer to an understanding of how to improve the quality of cheap wine and more research into this area is now called for as a matter of urgency.'

BAD ACADEMIA

What If Academics Were as Dumb as Quacks with Statistics?

Guardian, 10 September 2011

We all like to laugh at quacks when they misuse basic statistics. But what if academics, en masse, make mistakes that are equally foolish?

This week Sander Nieuwenhuis and colleagues publish a mighty torpedo in the journal *Nature Neuroscience*. They've identified one direct, stark statistical error that is so widespread it appears in about half of all the published papers surveyed from the academic neuroscience research literature.

To understand the scale of this problem, first we have to understand the statistical error they've identified. This will take four hundred words. At the end, you will understand an important aspect of statistics better than half the professional university academics currently publishing in the field of neuroscience.

Let's say you're working on some nerve cells, measuring the frequency with which they fire. When you drop a particular chemical on them, they seem to fire more slowly. You've got some normal mice, and some mutant mice. You want to see if their cells are differently affected by the chemical. So you measure the firing rate before and after applying the chemical, both in the mutant mice, and in the normal mice.

When you drop the chemical on the mutant-mice nerve cells, their firing rate drops by, let's say, 30 per cent. With the number of mice you have (in your imaginary experiment), this difference is statistically significant, which means it is unlikely to be due to chance. That's a useful finding which you can maybe publish. When you drop the chemical on the normal-mice nerve cells, there is a bit of a drop in firing rate, but not as much – let's say the drop is 15 per cent – and this smaller drop doesn't reach statistical significance.

But here is the catch. You can say that there is a statistically significant effect for your chemical reducing the firing rate in the mutant cells. And you can say there is *no* such statistically significant effect in the normal cells. But you cannot say that mutant cells and normal cells respond to the chemical differently. To say that, you would have to do a third statistical test, specifically comparing the 'difference in differences', the difference between the chemical-induced change in firing rate for the normal cells against the chemical-induced change in the mutant cells.

Now, looking at the figures I've given you here (entirely made up, for our made-up experiment), it's very likely that this 'difference in differences' would not be statistically significant, because the responses to the chemical only differ from each other by 15 per cent, and we saw earlier that a drop of 15 per cent on its own wasn't enough to achieve statistical significance.

But in exactly this situation, academics in neuroscience papers are routinely claiming that they have found a difference in response, in every field imaginable, with all kinds of stimuli and interventions: comparing responses in younger versus older participants; in patients against normal volunteers; in one task against another; between different brain areas; and so on.

How often? Nieuwenhuis and colleagues looked at 513 papers published in five prestigious neuroscience journals over two years. In half the 157 studies where this error could have been made, it was made. They broadened their search to 120 cellular and molecular articles in *Nature Neuroscience* during 2009 and 2010: they found twenty-five studies committing this statistical fallacy, and not one single paper analysed differences in effect sizes correctly.

These errors are appearing throughout the most prestigious journals in the field of neuroscience. How can we explain that? Analysing data correctly, to identify a 'difference in differences', is a little tricksy, so thinking very generously, we might suggest that researchers worry it's too long-winded for a paper, or too difficult for readers. Alternatively, perhaps less generously, we might decide it's too tricky for the researchers themselves.

But the darkest thought of all is this: analysing a 'difference in differences' properly is much less likely to give you a statistically significant result, and so it's much less likely to produce the kind of positive finding you need to get your study published, to get a point on your CV, to get claps at conferences, and to get a good feeling in your belly. In all seriousness: I hope this error is only being driven by incompetence.

Brain-Imaging Studies Report More Positive Findings Than Their Numbers Can Support. This Is Fishy

Guardian, 13 August 2011

While the authorities are distracted by mass disorder, we can do some statistics. You'll have seen plenty of news stories telling you that one part of the brain is bigger, or smaller, in people with a particular mental health problem, or even a specific job. These are generally based on real, published scientific research. But how reliable are the studies?

One way of critiquing a piece of research is to read the academic paper itself, in detail, looking for flaws. But that might not be enough, if some sources of bias might exist outside the paper, in the wider system of science.

By now you'll be familiar with publication bias: the phenomenon whereby studies with boring negative results are less likely to get written up, and less likely to get published. Normally you can estimate this using a tool such as, say, a funnel plot. The principle behind these is simple: big, expensive landmark studies are harder to brush under the carpet, but small studies can disappear more easily. So essentially you split your studies into 'big ones' and 'small ones': if the small studies, averaged out together, give a more positive result than the big studies, then maybe some small negative studies have gone missing in action.

Sadly, this doesn't work for brain-scan studies, because there's not enough variation in size. So Professor John Ioannidis, a godlike figure in the field of 'research about research', took a different approach. He collected a large representative sample of these anatomical studies, counted up how many positive results they got, and how positive those results were, and then compared this to how many similarly positive results you could plausibly have expected to detect, simply from the sizes of the studies.

This can be derived from something called the 'power calculation'. Everyone knows that bigger is better when collecting data for a piece of research: the more you have, the greater your ability to detect a modest effect. What people often miss is that the size of the sample needed also changes with the size of the effect you're trying to detect: detecting a true 0.2 per cent difference in the size of the hippocampus between two groups, say, would need more subjects than a study aiming to detect a huge 25 per cent difference.

By working backwards and sideways from these kinds of calculations, Ioannidis was able to determine, from the sizes of effects measured, and from the numbers of people scanned, how many positive findings could plausibly have been expected, and compare that to how many were actually reported. The answer was stark: even being generous, there were twice as many positive findings as you could realistically have expected from the amount of data reported on.

What could explain this? Inadequate blinding is an issue: a fair amount of judgement goes into measuring the size of a brain area on a scan, so wishful nudges can creep in. And boring old publication bias is another: maybe whole negative papers aren't getting published.

But a final, more interesting explanation is also possible. In these kinds of studies, it's possible that many brain areas are measured to see if they're bigger or smaller, and maybe then only the positive findings get reported *within* each study.

There is one final line of evidence to support this. In studies of depression, for example, thirty-one studies report data on the hippocampus, six on the putamen, and seven on the prefrontal cortex. Maybe, perhaps, more investigators really did focus solely on the hippocampus. But given how easy it is to measure the size of another area – once you've recruited and scanned your participants – it's also possible that people are measuring these other areas, finding no change, and not bothering to report that negative result in their paper alongside the positive ones they've found.

There's only one way to prevent this: researchers would have to publicly pre-register what areas they plan to measure, before they begin, and report all findings. In the absence of that process, the entire field might be distorted by a form of exaggeration that is – we trust – honest and unconscious, but more interestingly, collective and disseminated.

'None of Your Damn Business'

Guardian, 15 January 2011

Sometimes something will go wrong with an academic paper, and it will need to be retracted: that's entirely expected. What matters is how academic journals deal with problems when they arise.

In 2004 the *Annals of Thoracic Surgery* published a study comparing two heart drugs. This week it was retracted. Ivan Oransky and Adam Marcus are two geeks who set up a website called RetractionWatch because it was clear that retractions are often handled badly: they contacted the editor of *ATS*, Dr L. Henry Edmunds Jr, MD to find out why the paper was retracted. 'It's none of your damn business,' replied Dr Edmunds, before railing against 'journalists and bloggists'. The retraction notice, he said, was merely there 'to inform our readers that the article is retracted'. 'If you get divorced from your wife, the public doesn't need to know the details.'

ATS's retraction notice on this paper is equally uninformative and opaque. The paper was retracted 'following an investigation by the University of Florida, which uncovered instances of repetitious, tabulated data from previously published studies'. Does that mean duplicate publication, two bites of the cherry? Or maybe plagiarism? And if so, of what, by whom? And can we still trust the authors' numerous other papers?

What's odd is that this is not uncommon. Academic journals

have high expectations of academic authors, with explicit descriptions of every step in an experiment, clear references, peer review, declarations for financial conflicts of interest, and so on, for a good reason: academic journals are there to inform academics about the results of experiments, and to discuss their interpretation. Retractions form an important part of that record.

Here's one example of why. In October 2010 the *Journal of the American Chemical Society* retracted a 2009 paper about a new technique for measuring DNA, explaining it was because of 'inaccurate DNA hybridization detection results caused by application of an incorrect data processing method'. This tells you nothing. When RetractionWatch got in touch with the author, he explained that his team forgot to correct for something in their analysis, which made the technique they were testing appear to be more powerful than it really was; they actually found it's no better than the process it was proposed to replace.

That's useful information, much more informative than the paper simply disappearing one morning, and it clearly belongs in the academic journal the original paper appeared in, not in an email to two people from the internet running an ad hoc blog tracking down the stories behind retractions.

This all becomes especially important when you think through how academic papers are used: that *JACS* paper has now been cited fourteen times, by people who believed it to be true. And we know that news of even the simple fact of a retraction fails to permeate through to consumers of information.

Researcher Stephen Breuning faked huge amounts of trial data on the drug ritalin, and was found guilty of scientific misconduct in 1988 by a US federal judge – which is unusual and extreme in itself – so most of his papers were retracted. A study last year chased up all the references to Breuning's work from 1989 to 2007, and found over a dozen academic papers still citing his work. Some discussed it as a case of fraud, but around half – in more prominent journals – still cited it as if it was valid, twenty-four years after its retraction.

The role of journals in policing academic misconduct is still unclear, but obviously, explaining the disappearance of a paper you published is a bare minimum. Like publication bias, whereby negative findings are less likely to be published, this is a systemic failure, across all fields, so it has far greater ramifications than any one single, eye-catching academic cock-up or fraud. Unfortunately it's also a boring corner in the technical world of academia, so nobody has been shamed into fixing it. Eyeballs are an excellent disinfectant: you should read RetractionWatch.

Twelve Monkeys. No . . . Eight. Wait, Sorry, I Meant Fourteen

Guardian, 23 January 2010

Like many people, you're possibly afraid to share your views on animal experiments, because you don't want anyone digging up your grandmother's grave, or setting fire to your house, or stuff like that. Animal experiments are necessary, they need to be properly regulated, and we have some of the tightest regulation in the world.

But it's easy to assess whether animals are treated well, or whether an experiment was necessary. In the nerd corner there is another issue: is the research well conducted, and are the results properly communicated? If it's not, then animals have suffered – whatever you believe that might mean for an animal – partly in vain.

The National Centre for the Replacement, Refinement and Reduction of Animals in Research was set up by the government in 2004. It has published, in the academic journal *PLoS One*, a systematic survey of the quality of reporting, experimental design and statistical analysis of recently published biomedical research using laboratory animals. These results are not good news.

The study is pretty solid. It describes the strategy they used to search for papers, which is important, because you don't want to be like a homeopath, and only quote the papers that support your conclusions: you want to have a representative sample of all the literature. And the papers they found covered a huge range of publicly funded research: behavioural and diet studies, drug and chemical testing, immunological experiments, and more.

Some of the flaws they discovered were bizarre. Four per cent of papers didn't mention how many animals were used in the experiment, anywhere. The researchers looked in detail at forty-eight studies that did say how many were used: not one explained why that particular number of animals had been chosen. Thirty-five per cent of the papers gave one figure for the number of animals used in the methods, and then a different number of animals appeared in the results. That's pretty disorganised.

They looked at how many studies used basic strategies to reduce bias in their results, like randomisation and blinding. If you're comparing one intervention against another, for example, and you don't randomly assign animals to each group, then it's possible you might unconsciously put the stronger animals in the group getting a potentially beneficial experimental intervention, or vice versa, thus distorting your results.

If you don't 'blind', then you know, as the experimenter, which animals had which intervention. So you might allow that knowledge, even unconsciously, to affect close calls on measurements you take. Or maybe you'll accept a high blood-pressure reading when you expected it to be high, knowing what you do about your own experiment, but then double-check a high blood-pressure measurement in an animal where you expected it to be low.

Only 12 per cent of the animal studies used randomisation. Only 14 per cent used blinding. And the reporting was often poor. Only 8 per cent gave the raw data, allowing you to go back and do your own analysis. About half the studies left the numbers of animals in each group out of their tables.

I grew up friends with the daughters of Colin Blakemore, a neuroscientist in Oxford who has taken courageous risks over many decades to speak out and defend necessary animal research. My first kiss – not one of those sisters, I should say – was outside a teenage party in a church hall, in front of two Special Branch officers sitting in a car with their lights off.

People who threaten the lives of fifteen-year-old girls, to shut their father up, are beneath contempt. People who fail to damn these threats are similarly contemptible. That's why it sticks in the throat to say that the reporting and conduct of animal research is often poor; but we have to be better.

Medical Hypotheses Fails the Aids Test

Guardian, 12 September 2009

This week the peer-review system has been in the newspapers, after a survey of scientists suggested it had some problems. This is barely news. Peer review – where articles submitted to an academic journal are reviewed by other scientists from the same field for an opinion on their quality – has always been recognised as problematic. It is time-consuming, it can be open to corruption, and it cannot always prevent fraud, plagiarism or duplicate publication, although in a more obvious case it might. The main problem with peer review is: it's hard to find anything better.

Here is one example of a failing alternative. This month, after a concerted campaign by academics aggregating around websites such as Aidstruth.org, academic publishers Elsevier have withdrawn two papers from a journal called *Medical Hypotheses*. This academic journal is a rarity: it does not have peer review; instead, submissions are approved for publication by its one editor.

Articles from *Medical Hypotheses* have appeared in this column quite a lot. It carried one almost surreally crass paper* in which two Italian doctors argued that 'mongoloid' really was an appropriate term for people with Down's syndrome after all, because they share many characteristics with Oriental populations (including: sitting cross-legged, eating small amounts of lots of different types of food with MSG in it, and an enjoyment of handicrafts). You might also remember two pieces discussing the benefits and side effects of masturbation as a treatment for nasal congestion.*†

The papers withdrawn this month step into a new domain of foolishness. Both were from the community whose members characterise themselves as 'Aids dissidents', and one was co-authored by its figureheads, Peter Duesberg and David Rasnick.

To say that a peer reviewer might have spotted the flaws in their paper – which had already been rejected by the *Journal of Aids* – is an understatement. My favourite part is the whole

---

* See page 141.
† See page 143.

page they devote to arguing that there cannot be lots of people dying of Aids in South Africa, because the population of that country has grown over the past few years.

We might expect anyone to spot such poor reasoning – and only two days passed between this paper's submission and its acceptance – but they also misrepresent landmark papers from the literature on Aids research. Rasnick and Duesberg discuss antiretroviral medications, which have side effects, but which have stopped Aids being a death sentence, and attack the notion that their benefits outweigh the toxicity: 'Contrary to these claims,' they say, 'hundreds of American and British researchers jointly published a collaborative analysis in the *Lancet* in 2006, concluding that treatment of Aids patients with anti-viral drugs has "not translated into a decrease in mortality".'

This is a simple, flat, unambiguous misrepresentation of the *Lancet* paper to which they refer. Antiretroviral medications have repeatedly been shown to save lives in systematic reviews of large numbers of well-conducted randomised controlled trials. The *Lancet* paper they reference simply surveys the first decade of patients who received highly active antiretroviral therapy (HAART) – modern combinations of multiple anti-retroviral medications – to see if things had improved, and they had not. Patients receiving HAART in 2003 did no better than patients receiving HAART in 1995. This doesn't mean that HAART is no better than placebo. It means outcomes for people on HAART didn't improve over an eight-year period of their use. This would be obvious to anyone familiar with the papers, but also to anyone who thought to spend the time checking the evidence for an obviously improbable assertion.

What does all this tell us about peer review? The editor of *Medical Hypotheses*, Bruce Charlton, has repeatedly argued – very reasonably – that the academic world benefits from having journals with different editorial models, that peer review can censor provocative ideas, and that scientists should be free to pontificate in their internal professional literature. But there are blogs where Aids dissidents, or anyone, can pontificate wildly and to their colleagues: from journals we expect a little more.

Twenty academics and others have now written to Medline, requesting that *Medical Hypotheses* should be removed from its index. Aids denialism in South Africa has been responsible for the unnecessary deaths of an estimated 330,000 people. You can do peer review well, or badly. You can follow the single-editor model well, or foolishly. This article was plainly foolish.

Observations on the Classification of Idiots

Guardian, 18 August 2007

Every now and then something comes along which is so bonkers and so unhinged that it unmoors itself from all cultural anchoring points, and floats off into a baffling universe all of its own. I am an enthusiast for bad ideas, but nothing prepared me for this, in the academic journal *Medical Hypotheses*: an article called 'Down Subjects and Oriental Population Share Several Specific Attitudes and Characteristics'.

You'd be right to experience a shudder of nervousness at the title alone, since this is an academic journal, from 2007, and not 1866, when John Langdon Down wrote his classic 'Observations on the Ethnic Classification of Idiots'. That paper was the first to describe Down's syndrome (which Down called 'mongolism'), and in it the author explained that different forms of genetic disorder were in fact evolutionary regressions to what he viewed as the less advanced, non-white forms of humanity.

He described an Ethiopian form of 'idiot', a mongoloid form, and so on. Looking back, it reads as spectacularly offensive.

Now. People with Down's syndrome – who have three copies of chromosome 21, learning difficulties and other congenital health problems – do indeed look, to Westerners, a tiny bit like people from East Asia. This is because they have something called an 'epicanthic fold', a piece of skin that joins the upper part of the nose to the inner part of the eyebrow. It makes the eyes almond-shaped. You'll find epicanthic folds on faces from East Asia, South-East Asia, and some West Africans and Native Americans. People with Down's syndrome have various other incidental anatomical differences too, if you're interested, such as a single crease in their palm.

Flash forward to 2007 – I think that's where we are – to two Italian doctors. They offer their theory that the parallels between Down's syndrome and 'Oriental' people go beyond this fleeting facial similarity. What is the evidence they have amassed? I offer it almost in its totality.

One aspect, they say, is alimentary characteristics. 'Down subjects adore having several dishes displayed on the table, and have a propensity for food which is rich in monosodium glutamate.'

I, too, adore having several dishes displayed upon the table.

Two doctors, in an academic journal, in 2007, go on: 'The tendencies of Down subjects to carry out recreative-rehabilitative activities, such as embroidery, wicker-working, ceramics, bookbinding, etc., that is renowned, remind [us of] the Chinese handcrafts, which need a notable ability, such as Chinese vases, or the use of chopsticks employed for eating by Asiatic populations.'

Perhaps you can think of cultural rather than genetic explanations for these observations.

There's more. 'Down persons during waiting periods, when they get tired of standing up straight, crouch, squatting down, reminding us of the "squatting" position . . . They remain in this position for several minutes and only to rest themselves.' Amazing. 'This position is the same taken by the Vietnamese, the Thai, the Cambodian, the Chinese, while they are waiting at a bus stop, for instance, or while they are chatting.'

And that's not all. 'There is another pose taken by Down subjects while they are sitting on a chair: they sit with their legs crossed while they are eating, writing, watching TV, as the Oriental peoples do.'

To me – and I may be wrong – this article is so fantastical, so ridiculous, and so thoughtlessly crass, that it's hard to experience anything like outrage. But it appears in a proper academic journal, published by Elsevier, with a respectable 'impact factor' – a measure of how frequently a journal is cited – of 1.299. I contacted the editor. He told me the paper was a very short, discursive and preliminary communication, floating a general idea for discussion and debate, and that taking scientific ideas out of their context could be misleading. I hope I am not misleading anybody. I contacted Elsevier, the journal publisher: they will consider making the article free to access, so that anyone can read it for themselves. You can reach your own conclusions.

More Crap Journals?

Guardian, 4 October 2008

Important and timely news from the journal *Medical Hypotheses* this week: ejaculating could be 'a potential treatment of nasal congestion in mature males'. My reason for bothering you with this will become clear later.

The first thing to note is that this is not an entirely ludicrous idea, but it is a tenuous one. Most decongestant pills work by increasing the activity in something called the 'sympathetic nervous system', which is involved in lots of largely automatic things in the body, like sweating, blood pressure and pupil size, as well as the 'fight or flight' mechanism. More activity in the sympathetic system causes the vessels of the nasal mucosa to constrict, reducing their volume and so clearing the blockage; but these pills can also have lots of fairly unpleasant side effects, because they tend to affect the whole of the sympathetic nervous system.

The argument from Dr Zarrintan is as follows. 'The emission phase of ejaculation is under the control of the sympathetic nervous system . . . ejaculation will stimulate adrenergic receptors . . . and stimulation of your adrenergic receptors will give you relief from your cold.' It's a chain of reasoning that would make a nutritionist blush, and it has already been responded to by a letter entitled 'Ejaculation as a treatment for nasal congestion in men is inconvenient, unreliable and potentially hazardous'. This response explains that ejaculation increases blood pressure and heart rate, which has its own side effects, increases androgens in the body which could increase prostate cancer, and so on. I honestly don't know who's kidding any more.

Now, I genuinely love *Medical Hypotheses*, published by Elsevier. Last year, you will remember, it carried an almost surreally crass paper in which two Italian doctors argued that 'mongoloid' really was an appropriate term for people with Down's syndrome after all, because such people share many characteristics with Oriental populations.

Its articles are routinely quoted with great authority in the output of anti-vaccination conspiracy theorists, miracle-cure marketers and other interesting characters, but it also prints some interesting stuff. In that sense it serves a useful purpose, but it also acts as an extreme example of something we should all be aware of: you're not supposed to take everything in an academic journal as read, final and valid.

I once had a conversation with *Medical Hypotheses'* editor, Dr Bruce Charlton, and he raised two excellent points on the value of publishing loopy papers (that's my phrasing – you can read more from him online). The first was that academics must be free to simply get on and publish things that outsiders might find weird, or misinterpret, without worrying about what the wider public might think.

The Downs paper above was simply uninformative and offensive, pushing this argument to the limit, but excepting such cases, his is a view I would heartily endorse. Academics should be free to write tenuous papers. The infamous 1998 *Lancet* MMR paper is a perfect example. It described the experiences of twelve children with autism and some bowel problems, who'd had the MMR vaccine. This didn't tell us much about the chances of MMR causing autism. But nobody should censor themselves from publishing such work, that might be of tenuous use or interest to somebody somewhere, on the off-chance that doing so might trigger a ten-year-long epic scare story from mischievous journalists. (We now know, much later, that the contents of that *Lancet* article were themselves the result of scientific misconduct; this is a separate issue.)

But Charlton also raises a more interesting point. He feels that the ideas market requires a diverse range of publication venues, so his journal is deliberately not 'peer-reviewed': the process whereby the great and the good look at your article and decide if it is worth publishing, or is methodologically flawed. Peer review is a system that has worked OK, to an extent, to stop outright nonsense appearing in very competitive high-quality journals; but it is also riddled with holes, it acts as no bar to nonsense being published in obscure peer-reviewed journals (where the bar is much lower), and it's vulnerable to bullying and corruption.

Charlton's journal publishes ideas rather than data. But we have to accept that a large amount of bad-quality data is being published in the 5,000 medical academic journals that already exist (printing fifteen million papers to date), and in many respects we have to hope that this situation will get even worse. In a recent column I described how only one in four cancer trials is actually published. There are widespread demands that all negative findings must be published, so that they are at least accessible, but this will often mean that inadequately analysed data from less competent studies are placed in repositories, or published in journals that will take very poor-quality papers.

The signal-to-noise ratio in the scientific literature is getting ever lower, and the simple fact that something has been 'published' is losing its currency as a badge of quality. That may, paradoxically, not be a bad thing. Academic papers are filled with ideas and evidence to be read, weighed up, and critically appraised, by people with the motivation and skills to do so, whoever they may be. Science is not, and should not be, about arguing from authority. The idea that the conclusions of a published paper are automatically true was never helpful. The academic literature is a buyer-beware environment.

GOVERNMENT STATISTICS

If You Want to Be Trusted More: Claim Less

Guardian, 8 January 2009

'Public Sector Pay Races Ahead in a Recession', shouted the front page of this week's *Sunday Times*. 'Public sector workers earn 7 per cent more on average than their peers in the private sector – a pay gulf that has more than doubled since the recession began.' The *Telegraph* followed up with a copycat story a few hours later.

In reality, this is one of those interesting areas where anybody who makes a firm statement is wrong, because there is not sufficient evidence to make a confident assertion in either direction.

The *Sunday Times* has identified a difference in the median pay of all public sector employees in the country, when compared with all the private sector employees in the country. It has then over-extrapolated from these two figures to claim that – job for job – public sector employees are paid more than their peers in the private sector.

We will discuss why that analysis is worse than useless in a moment.

But first, some interesting details. For its analysis the *Sunday Times* uses 'annual salary' instead of 'hourly pay', although the latter is clearly more meaningful, especially since the newspaper quotes the annual salary figures for part-time and full-time employees, all mixed together, but 31 per cent of public sector jobs are part-time, against 23 per cent of private sector jobs. In fact, quoting 'hourly salary' would also have made the difference between the public and private sector median wages look even bigger. So why did the *Sunday Times* and the *Telegraph* use annual pay?

Perhaps because this figure makes the difference in medians look like a new phenomenon under the present Labour government. Using the hourly figures, you can see that public sector median hourly pay has been higher than private sector pay for years. If you go to the 'Annual Survey of Hours and Earnings' data on the ONS website which the *Sunday Times* used, you can see for yourself. It was £7.98 vs £6.72 in 1997 under the previous (Conservative) government, a difference of almost 20 per cent, and £8.56 vs £7.32 in 1999. Meanwhile, the 'annual salary' difference which the *Sunday Times* chose to use was negligible in 1999 (the first year ONS gave this figure), at £15,002 vs £14,963, a difference of 0.3 per cent, allowing the paper to create the illusion of a brand-new phenomenon:

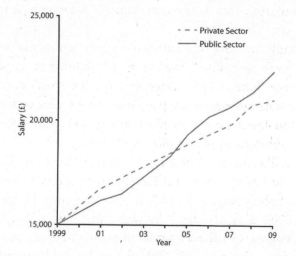

More than that, using the 'annual salary' figure allows the *Sunday Times* to claim dramatically that the difference has almost doubled in two years: the difference in medians for annual pay has gone from 3.8 per cent to 6.8 per cent since 2007, while the difference in hourly pay has gone from 25.1 per cent to 28.7 per cent, which is much less eye-catching.

'By a whole range of measures,' the *Sunday Times* continues, 'public sector employees are also enjoying better working conditions. Last year the average public sector worker laboured for 35 hours a week . . . 2 hours less than the typical private sector worker.'

Is this really down to laziness, and better working conditions? No. Again, this is simply due to the greater number of part-time jobs in the public sector – 31 per cent vs 23 per cent – which is a long-standing phenomenon.

But there is a deeper problem with the analysis in the *Sunday Times* and the *Telegraph*. The long-standing difference in median wage for all jobs in each sector is hardly informative on the question of whether someone is paid more or less than their peer in the other sector. Firstly, it's hard to decide what the comparison job is for a policeman, a fireman, a teacher, and so on.

Secondly, to make that comparison between medians meaningful, you'd need data showing the breakdown of what kinds of jobs are done in each sector. Because it's possible, after all, that the state employs more people in more senior or middling roles, and fewer people in the kinds of jobs you find at the absolute bottom of the employment ladder.

If you like, for an illustration, we can poke around the ONS Annual Survey of Hours and Earnings data again. The national median hourly wage is £11.03. If you take table 14_5a of the ASHE 2009 data, reorder it by wage, and look at the bottom three categories with over a million people in them as a rough illustration, we have: 1,126,000 sales and retail assistants on a median hourly wage of £6.36; 1,355,000 cashiers at £6.40; 1,430,000 in sales at £6.45.

None of these are jobs you find in the public sector, although there are also cleaners at the low-wage end of this table. If someone here was quoting data comparing public/private wages for the same kind of cleaning jobs, say, then that would be interesting. There's no such data on offer. But as the *Sunday Times* says: 'Our reports today show, the public sector has become so big and such a generous employer that it is sucking workers out of private companies.' I don't see how it can justify this, other than with its own laughable case studies, and if it's true, it should be a long-standing trend, not a new one.

I could go on. It's not surprising if public sector pay increased from what it used to be under this government: improving recruitment for teachers and the like was a manifesto promise. But as for a comparison, I don't know if the public sector pays more than the private sector for the same work, or less: nobody does, from a difference in median wages. Meanwhile I do know that this was one of the most statistically misleading front-page stories I have seen in a long time. It's going to be a fun election.

Is This the Worst Government Statistic Ever Created?

Guardian, 24 June 2011

Every now and then, the government will push a report that's so asinine, and so thin, you have to check it's not a spoof. The *Daily Mail* was clear in its coverage: 'Council incompetence "costs every household £452 a year"'; 'Up to £10bn a year is wasted by clueless councils'. And the *Express* agreed. Where will this money come from? 'Up to £10 billion a year could be saved . . . if councils better analysed spending from their £50 billion procurement budgets.'

A 20 per cent saving on the £50 billion council procurement budget would be awesome. And this is a proper story, from a press release on the Department for Communities and Local Government website: 20 per cent of the £50 billion procurement spend could be saved by seeking better value.

Government ministers have an army of intelligent technical staff, with full access to every speck of data, ready to produce research. But these figures come from a 'new, cutting-edge analysis of council spending data by procurement experts Opera Solutions'.

I downloaded the 'Opera Solutions White Paper'. I recommend reading it yourself, to understand what a minister considers a substantive piece of research.

The 'full report' is six pages long, not including the cover. The meat of it, the analysis, is presented in a single three-line table. Opera took the recently released local government spending data for three councils, and decided how much it reckoned could be saved by bulk purchasing.

It did its estimates on three areas: for energy bills (a £7 million spend) and solicitors' fees (£6 million), it thought councils could save just 10 per cent. The third category – mobile-phone bills – was tiny in comparison (just £600,000 spent), but here, and here alone, Opera reckons councils can save 20 per cent by getting people on better tariffs.

So, for mobile phones, an incompetently regulated sector well known for making money from deliberately confusing pricing schemes, where phone companies hope customers will regard checking their usage and changing tariffs as more effort than it's worth, Opera reckons councils can save 20 per cent.

Then, even though for £13 million out of £13.6 million of its spend calculations Opera could only find 10 per cent of savings, it cheerfully applies this magic 20 per cent from the tiny mobile-phone spend to the entire local government procurement budget of £50 billion, magicking up £10 billion of savings, £452 a year for every one of us.

And even before that astonishing, shameless bait and switch, these figures are all presented out of nowhere. There is no working at all for any single saving, no description of how 10 per cent or even 20 per cent was calculated: just that three-line table telling you how much Opera Solutions *reckons* councils can save. There's also no justification for choosing energy, solicitors and mobile-phone bills, out of all the things councils spend on. Were these where Opera thought it could get the biggest savings? Who knows.

The document is six pages long. We've covered one page. What's in the rest? All that follows is a four-page glossy brochure advert for Opera Solutions' management consultancy services in local government. 'Opera Solutions has successfully completed procurement optimisation projects for hundreds of organisations around the world'; 'Opera partners with clients to work as a catalyst'; 'Opera addresses these issues through Insight Cube™ technology, which creates deep visibility into spending information.'

Meanwhile, back in the real world, what do local governments actually procure? Well, the biggest thing, about a quarter of that £50 billion budget, more than £10 billion a year of local government procurement, is social care: mostly residential care, mostly for the elderly, and most through the independent sector.

If you're going to save 20 per cent off that, then I suggest you tell us how, in full and educative detail. In the meantime, saying you can get us a better deal on our mobile-phone tariff, and then pretending that means you've taken 20 per cent off the entire £50 billion local government procurement spend, isn't just misleading: it's the reasoning of a ten-year-old.

Anarchy for the UK. Ish.

Guardian, 2 April 2011

Here are two fun ways that numbers can be distorted for political purposes. Each of them feels oddly poetic in its ability to smear or stifle.

The first is simple: you can conflate two different things into one omnibus figure, either to inflate a problem, or to confuse it. Last weekend a few hundred thousand people marched in London against government cuts. On the same day there was some violent disturbance, windows smashed, policemen injured, and drunkenness.

The *Sun* said: 'Police have charged nearly 150 people after violent anarchists hijacked the anti-cuts demo and brought terror to London's streets.' The *Guardian* republished a Press Association report headlined 'Cuts protest violence: 149 people charged'. And from the locals, for example, the *Manchester Evening News* carried 'Boy, 17, from Manchester Among 149 Charged Over Violence After Anti-Cuts March'.

In reality, a dozen of these charges related to violence, while 138 were people who were involved in an apparently peaceful occupation of Fortnum & Mason organised by UKUncut, who campaign against tax avoidance.

You will have your own view on whether people should be arrested and charged for standing in a shop as an act of protest. But describing these 150 people as 'violent anarchists . . . who brought terror to London's streets' is not just misleading; it also makes the police look over twelve times more effective than they really were at charging people who perpetrated acts of violence.

The second method of obfuscation is even simpler. After London was chosen to host the 2012 Olympics, Labour made a series of pledges, including two around health: to use the power of the Games to inspire a million more people to play sport three or more times a week; and to get a million more people doing more general physical activity.

Politicians seem keen on the idea that large multi-sports events can have a positive impact like this, so the area has been studied fairly frequently. Last year the *BMJ* published a systematic review of the literature. It set out to find any study that had ever been conducted into the real-world health and socio-economic impacts of major multi-sport events on the host population.

This research found fifty-four studies. Overall, the quality was poor (it's a fairly difficult thing to measure, and most studies used cross-sectional surveys, repeated over time). The bottom line was this: there is no evidence that events like the Olympics have a positive impact on either health or socio-economic outcomes.

Here are some examples from the review. One study looked at Manchester before and after the 2002 Commonwealth Games: overall sports participation (four times or more in the past month) fell after the Games, and the gap in participation rates between rich and poor areas widened significantly.

Another study in Manchester suggested there were particular problems around voluntary groups being excluded from using Commonwealth branding, and that new facilities tended to benefit elite athletes rather than the general population.

There was a vague upward trend in sports participation in Barcelona between the early 1980s and 1994, and it had the Olympics in 1992. Volunteers working at the Commonwealth Games showed no increase in sports participation.

You will have your own views on whether the cost of hosting the Olympics is proportionate to the benefits, and where those benefits lie. From this systematic review, however, there's no evidence for large multi-sports events having a positive health or socioeconomic impact overall, so only an optimist would make promises to the contrary.

This week, it emerged that both of the government's targets for improving healthy activity after the 2012 Olympics are now being quietly dropped. By walking away from outcome indicators that will not be met, a government can create a false impression of success: if prespecified outcome indicators are ever to mean anything, after all, it's because you report on all of them clearly, whether success is achieved or not.

But more than that, governments around the world spend billions of pounds on these events. By quietly dropping these outcome indicators, rather than carefully documenting our success or failure at meeting them, our current politicians pave the way for ever more false and over-optimistic claims by their colleagues, all around the world, for many years to come.

More Than Sixty Children Saved from Abuse

Guardian, 7 August 2010

According to the Home Office this week, Sarah's Law – by which any parent can find out if any adult in contact with their child has a record of violent or sexual crimes – has 'already protected more than 60 children from abuse during its pilot'. This fact was widely reported, and was the headline finding. As the *Sun* said: 'More than sixty sickening offences were halted by Sarah's Law during its trial.'

It seems to me that the number of sickening offences prevented by an intervention is a difficult thing to calculate:

nobody explained where the number came from, so for my own interest I called the Home Office.

'It's not that difficult to work out is it?' This is the Home Office telling me I'm stupid. 'It's the number of disclosures issued, how many were of sex offenders, and how many children would those offenders have had contact with.' This means telling a parent that someone in contact with their child had a history of abuse equated to preventing an act of abuse? Yes, they said: 'Protecting that child means ensuring that offender did not have a way of having contact with that child. Therefore that child is being protected.' This assumes that any such contact is itself abusive, or would definitely result in abuse. That might be correct: I slightly doubt it, but I don't know for sure.

Then I asked where the number sixty came from. I was sent to an excellent report assessing the programme, written by a team of academics. Neither the number 60 nor the word sixty appears in that document.

So I contacted the lead author, Prof Hazel Kemshall, who said: 'You are correct that reference to sixty children is not made in the report. As I understand it the Home Office have drawn on police data sources to quote this figure, and therefore I cannot assist you further. As you will see from the report, we were careful to state the limits of the methodology.'

I contacted the Home Office again. 'The figure is over sixty and it comes from the number of disclosures made where there was a conviction of a sexual offence with a minor or violence against a minor. In total twenty-one disclosures were made specifically about registered sex offenders (RSO), a further eleven disclosures were made, for example relating to convictions for violent offending. These people had access to over sixty children.'

I'm not sure this is self-evident. The academics who wrote the report couldn't work out where the number sixty came from, and at least two pieces have appeared trying to unpick it, each coming up with different answers from me and the Home Office. The excellently named Conrad Quilty-Harper in the *Telegraph* and a promising new website called FullFact, both – very reasonably – tried adding various categories of numbers from the academic report, including a figure on social-worker activity which seemed to make up the numbers.

I'm not saying the figure sixty is wrong. While what it represents was probably overstated – by the Home Office and the press reports – the number itself isn't absurd. But it does seem odd that just finding out where it came from involved so much mucking about, and it seems even odder to ignore the robust figures in a long academic report that you've commissioned (the scheme wasn't cheap compared to, say, social-worker salaries), and instead build your press activity around one opaque figure constructed, ever so slightly, somewhere, it seems, on the back of an envelope.

Home Taping Didn't Kill Music

Guardian, 6 June 2009

You are killing our creative industries. 'Downloading Costs Billions', said the *Sun*. 'MORE than seven million Brits use illegal downloading sites that cost the economy billions of pounds, Government advisors said today. Researchers found more than a million people using a download site in ONE day and estimated that in a year they would use £120 billion worth of material.'

That's about a tenth of our GDP. No wonder the *Daily Mail* was worried too: 'The network had 1.3 million users sharing files online at midday on a weekday. If each of those downloaded just one file per day, this would amount to 4.73 billion items being consumed for free every year.'

Now, I am always suspicious of anything on piracy from the music industry, because it has produced a lot of dodgy figures over the years. I also doubt that every download is lost revenue, since, for example, people who download more music also buy more music. I'd like more details.

So where do these notions of so many billions in lost revenue come from? I found the original report. It was written by some academics you can hire in a unit at UCL called CIBER, the Centre for Information Behaviour and the Evaluation of Research (which 'seeks to inform by countering idle speculation and uninformed opinion with the facts'). The report was commissioned by a government body called SABIP, the Strategic Advisory Board for Intellectual Property Policy.

On the billions lost it says: 'Estimates as to the overall lost revenues if we include all creative industries whose products can be copied digitally, or counterfeited, reach £10 billion (IP Rights, 2004), conservatively, as our figure is from 2004, and a loss of 4,000 jobs.'

What is the origin of this conservative figure? I hunted down the full CIBER documents, found the references section, and followed the web link, which led to a 2004 press release from a private legal firm called Rouse which specialises in intellectual property law. This press release was not about the £10 billion figure. It was, in fact, a one-page document which simply welcomed the government setting up an intellectual property theft strategy. In a short section headed 'Background', among five other points, it says: 'Rights owners have estimated that last year alone counterfeiting and piracy cost the UK economy £10 billion and 4,000 jobs.' So this authoritative government figure, from an academic study, in fact comes from an industry estimate, made as an aside, five years earlier, in a short press release from one law firm.

But what about all those other figures in the media coverage? Lots of it revolved around the figure of 4.73 billion items downloaded each year, worth £120 billion. This means each downloaded item – software, movie, mp3, ebook – is worth £25 on average. Now, before we go anywhere, this seems very high. I am not an economist, and I don't know about their methods, but to me, an appropriate comparator for someone who downloads a film to watch it once might be the rental value, for example, not the sale value. I'd also like to suggest that sometimes, perhaps quite often, someone who downloads a £1,000 professional 3D animation software package to fiddle about with at home may not use it more than three times.

In any case, that's £175 a week, or £9,100 a year, potentially not being spent by millions of people. Is this really lost revenue for the economy, as reported in the press? Plenty of those downloading will have been schoolkids, or students, who may not have £9,100 a year to spend. Even if they weren't, that figure is still about a third of the average UK wage. Before tax. Oh, and the government's figures were wrong: it was actually 473 million items, and £12 billion, so the value of each downloaded item was still £25, but it exaggerated the amount of money 'lost' by a factor of ten, in the original executive summary, and in the press release. These were changed quietly after the errors were pointed out by a BBC journalist, but I can find no public correction for the many people who were misled.

So I asked SABIP what steps they took to notify journalists of their error, which resulted in their absurdly huge claims being widely reported in news outlets around the world. They refused to answer my questions in emails; insisted on a phone call (always a warning sign); told me that they had taken steps, but wouldn't say what; explained something about how they couldn't be held responsible for lazy journalism; then, bizarrely, after ten minutes, tried to tell me retrospectively that the whole call was actually off the record, that I wasn't allowed to use the information in my piece, but that they had answered my questions, and so they didn't need to answer on the record, but I wasn't allowed to use the answers, and I couldn't say they hadn't answered, I just couldn't say what the answers were. Then the PR man from SABIP demanded that I acknowledge, in our phone call, formally, for reasons I still don't fully understand, that he had been helpful.

I think it's OK to be confused and disappointed by this. Like I said: as far as I'm concerned, everything from this industry is false, until proven otherwise.

Is This a Joke?

Guardian, 18 July 2009

We'd all like to help the police do their job well. They, in turn, would like to have a massive database with DNA profiles of everyone who has been arrested, but not convicted of a crime. We worry that this is intrusive, but some of us are willing to make concessions – on our principles and the invasion of our privacy – in the name of preventing crimes. To do this, we'd like to know the evidence on whether this database is helpful, to help us make an informed decision.

Luckily, the Home Office has now published a consultation paper on the subject. It defends the database by arguing that innocent people who have been arrested and released are basically criminals anyway, and go on to commit crimes in the future as much as guilty people do. 'This,' it says, 'is obviously a controversial assertion.' There is no reason for this assertion to be controversial: it's a simple factual matter, and if it's true, then you could easily assemble some good-quality evidence to prove it.

The Home Office has assembled some evidence. This study, from the Jill Dando Institute, attached to the consultation paper as an appendix, is possibly the most unclear and badly presented piece of research I have ever seen in a professional environment.

They want to show that the level of criminal activity in a group of people who have been arrested, but against whom no further action has been taken, is the same as the level of criminal activity in people who have been arrested and convicted of a crime, or who have accepted a caution.

On page 30 they explain their methods, haphazardly, scattered about in the text. They describe some people 'sampled on 1st June 2004, 1st June 2005 and 1st June 2006'. These dates are never mentioned again. I have no idea what their plan was there. They then leap to talking about Table 2. This contains data on people, each from a 'sample' in 1996, 1995 and 1994, followed up for thirty months, forty-two months and fifty-four months respectively. Are these anything to do with the people from 2004, 2005 and 2006? I have no idea, and it is impossible to tell.

In fact, I have no idea what 'sample' means – perhaps that was the date on which they were first arrested. I don't know why they were only followed up for thirty, forty-two and fifty-four months, instead of all the way from the 1990s to 2009. Crucially, I also don't know what the numbers in the table mean, because that isn't properly explained. I think it's the number of people from the original group who have subsequently been arrested again, but there's no way to tell.

Then they start to discuss the results from this table. They say that these figures show that arrested non-convicted people are the same as convicted people. There are no statistics conducted on these figures, so there is absolutely no indication of how wide the error margins are, and whether these are chance findings. To give you a hint about the impact of noise on their data, more people are described as having been subsequently re-arrested over the forty-two-month follow-up period than over the fifty-four-month follow-up period, which seems surprising, given that the people in the fifty-four-month group had a much longer period of time in which to get arrested.

This is before we even get on to the other problems. At a few hundred people, this study seems pretty small for one that is supposed to give compelling evidence that there is *no* difference between two groups, because to prove a negative like this, you generally need a very large sample, to minimise the chance of missing a true difference in the noise.

There is no evidence that they have done a 'power calculation' to determine the sample size they'd need, and in any case, their comparison group feels a bit rigged to me. In their 'convicted' sample they only count people who had a non-custodial sentence, and exclude people who got a custodial sentence, on the grounds that those people would be incapable of committing a crime during their incarceration. This also has the effect, however, of making the 'criminal' group really not very criminal, and so consequently a bit more likely to be similar to innocent people.

I could go on. Table 1 is so thoroughly 'not as described' as to be uninterpretable. In the text they talk about different cells on the table which are 'solid red', 'stippled yellow', and 'blank', when in fact the whole thing is just blue.

This research is incomprehensible and unreadable. Anybody who claims to have been persuaded by the data quoted here is telling you, loudly and clearly in the subtitles, that they don't need to understand – or possibly even read – a piece of research in order to find it compelling. People like this are not to be trusted, and if research of this calibre is what guides our policy on huge intrusions into the personal privacy of millions of innocent people, then they might as well be channelling spirits.

EVIDENCE-BASED POLICY

I'd Expect This from UKIP, or the *Daily Mail*. Not from a Government Leaflet

Guardian, 15 April 2011

The government has issued a new leaflet aiming to justify the latest round of redisorganisation in the NHS. This leaflet is called 'Working Together For A Stronger NHS'. It was produced by Number 10, it appears on the Department of Health website, and many of the figures it contains are misleading, out of date, or factually incorrect.

The leaflet begins, like much pseudoscience, with some uncontroversial truths: the number of people aged over eighty-five will double, and the cost of drugs is rising. This is all true.

Then comes the trouble. In large letters, alone on one entire page, you see: 'If the NHS was performing at truly world-class levels we would save an extra 5,000 lives from cancer every year.' The reference for this is a paper in the *British Journal of Cancer* called 'What if Cancer Survival in Britain were the Same as in Europe: How Many Deaths are Avoidable?'

This study does not aim to predict the future: in fact, it looks at data from 1985 to 1999 – seriously – which is a very long time ago. It finds that if we'd had the same cancer survival rates – more than twenty years ago, in the eighties and nineties – as the average in the EU, then we'd have had 7,000 fewer deaths per year. Not 5,000 fewer. To put the big number in context, by this study's calculation 6–7 per cent of UK cancer deaths in the 1990s were avoidable. Since then we've seen the massive 2000 NHS Cancer Plan, a new decade, and a new century. This paper says nothing about the number of lives we 'would save' each year by 2011, and citing it in that context is entirely misleading.

The next interesting figure misleads about a trend (we've seen this a lot from health ministers recently) and attempts to take the credit for a long-standing change. The leaflet says: 'Since May 2010 the NHS has gained 2,550 more doctors and has 3,000 fewer managers.' This is correct: full-time equivalent figures (my favourite) from NHS workforce data show 97,720 doctors in May 2010, and 100,197 in December. But NHS Information Centre figures show that between 1999 and 2009 the total number of doctors increased from 88,693 to 132,683, GPs from 28,354 to 36,085, and consultants from 21,410 to 34,654. Doctors take a while to grow, and they've been growing in number for a good long while.

Then we have choice: '95 per cent want more choice over their healthcare'. The source given is the twenty-fifth British Social Attitudes Survey. Interestingly, the government has just announced that it's going to stop funding the health questions in the British Social Attitudes Survey, so this valuable resource won't be around for long. The data was collected in 2007, it's free to download, and if you do so you'll see it didn't ask about 'more choice'. Question 583 asks how much choice you think NHS patients *currently* have ('a lot', 'a little', and so on), and Question 584 asks how much choice you think they *should* have.

How those responses were aggregated to get 95 per cent of people in favour of 'more choice' – a key justification for reform – is a mystery: many people will have said they have 'a little' choice, and that they should have 'a little' choice (we can't see how many from the aggregated data in tables of course; we need respondent-level data, because that's the only way we can link together an individual's responses to a sequence of questions). I asked the government how it produced its figure, since BSA25 doesn't have data on the question 'Would you like *more* choice?' I was told the source was table 3.1 in Chapter 3 of a book that costs £52, called *Do People Want More Choice and Diversity of Provision in Public Services.*

I got that book: it's the same old BSA25 data. It doesn't contain anything on 'more choice', and they got the title wrong: it's not *Do People Want More Choice . . .* It's *Do People Want Choice . . .* which shows how misleading they're being, and wasted me a lot of time in the library. None of this should be difficult. The facts in this plainly political pamphlet should be clean, correct, transparent and justified. As the government defies all reason by claiming that NHS staff support its reforms, we can only fear the results of its new 'listening exercise'.

Andrew Lansley and His Imaginary Evidence

Guardian, 5 February 2011

I have never heard one politician use the word 'evidence' so persistently, and so misleadingly, as Health Secretary Andrew Lansley defending his NHS reforms. Since he repeatedly claims that the evidence supports his plan, let's skim through what we can find on whether GP consortiums work, the benefits of competition, and the failures of the NHS.

Are GP consortiums better than Primary Care Trusts for

commissioning? There have been fifteen major reorganisations of the NHS in thirty years. We've had GP fundholders, GP multifunds, primary care groups, primary care trusts, family practitioner committees, purchasing consortiums, and more. After all this change, lots of data should have been gathered on the impact of specific strategies.

In reality, few of them were properly studied. Here are four papers on GP fundholding, which is broadly similar to Lansley's GP consortiums. Kay in 2002 found it was introduced and then abolished without any evidence of its effects. In 2006 Greener and Mannion found a mix of good and bad but no evidence that it improved patient care. In 1995 Coulter found nothing but gaps in the evidence and no evidence of any improvement in efficiency, responsiveness, or quality. Petchley found there was insufficient data to make any judgement. Lansley says he is following the evidence. I see no evidence to follow here.

Next, competition. Andrew Lansley has repeatedly denied that he is introducing competition on price. This is disturbing behaviour: his Bill explicitly introduces price-based competition – it's in paragraph 5:43 of his 'NHS Operating Framework'.

Does variable-price competition work in healthcare markets? Working from first principles, markets for healthcare in which people compete on price as well as quality might be expected to produce lower-quality healthcare, because prices are easy to measure, while quality, in healthcare, is surprisingly difficult to measure: so quality suffers.

It's hard to research this kind of thing, but even the evidence on *fixed-price* competition – where you compete on quality – is mixed. There are various ways to assess it. Often people choose an outcome – like the number of people who survive a heart attack – and compare this outcome in areas of more intense or less intense competition. Sometimes competition makes things worse, sometimes better.

For *variable-price* competition, which is what we're facing, things don't look good. Its introduction in New Jersey in the 1990s was associated with a worsening death rate from heart attacks, while in the UK, stopping variable-price competition was associated with improvement. These aren't clean, easy interventions to assess, but despite his using the word repeatedly, again the 'evidence' does not support Lansley here.

Lastly, there is the justification for reform. Both Lansley and David Cameron are – rather shamefully – overstating our mortality figures, in order to claim that the NHS is failing. Everyone wants more improvement, but money or a structural change does not produce an immediate and visible reduction in mortality from one thing, so it's hard to use these figures to pin blame or credit on anyone; interventions take time to have an impact, especially on things that kill you slowly; and NHS treatment isn't the only factor affecting how many people die of something. But since you're interested, to take just two things: mortality from cancer has fallen every year since 1995, and heart attack deaths have halved since 1997.

The government claims that our rate of death from heart attacks is double that in France, even though we spend the same on health. Health economist John Appleby instantly debunked this claim in the *BMJ*, and his piece will become a citation classic. From static 2006 figures in isolation, the government is right. But the trajectory of improvement in the UK is so phenomenal that if the straight line continues – as it has done for thirty years – we will be better than France within a year.

I'm not in favour of – or against – anything here: genuinely, all health-service administrative models baffle and bore me equally. But when Andrew Lansley says all the evidence supports his interventions, as he has done repeatedly, he is simply wrong. His wrongness is not a matter of opinion, it is a fact, and his pretence at data-driven faux neutrality is not just irritating, it's also hard to admire. There's no need to hide behind a cloak of scientific authority, murmuring the word 'evidence' into micro-phones. If your reforms are a matter of ideology, legacy, whim and faith, then like many of your predecessors you could simply say so, and leave 'evidence' to people who mean it.

Why Is Evidence So Hard for Politicians?

Guardian, 12 February 2011

One thing you hope for, with politicians, is that they won't make the same mistakes over and over again.

Last week we saw that the government has overstated the prob-lems in the NHS by using dodgy figures (to be precise, it used misleading static figures instead of time trends). We also saw that Andrew Lansley's frequently repeated claim – that his reforms are justified by evidence – is untrue: the evidence doesn't show that price-based competition improves outcomes (if anything, it makes things worse); and the evidence also doesn't show that GP consor-tiums improve outcomes (unless you cherry-pick only the positive findings). It's OK if your reforms aren't supported by existing evidence: you just shouldn't claim that they are.

Now Lansley's Junior Minister Paul Burstow MP has kindly responded, repeating the exact same mistakes again, only more clumsily. From a Minister, this is frightening, so we should see how he does it.

Burstow's letter:

In Ben Goldacre's pursuit of the evidence for NHS modernisa-tion ('Evidence? What evidence?', 5 February), he appears to have overlooked the impact assessment we published alongside the health and social care bill, where we present a thorough analysis of the evidence for and against our plans. As he will see, studies show that GP fundholding and practice-based commissioning delivered shorter waits and fewer referrals to hospitals for patients. The evidence on competition demonstrates that when it is well-designed and conducted on the basis of quality (as we are proposing), rather than price, it can drive up quality and efficiency.

We have not sought to understate the achievements of the NHS – but a 2008 study by Martin McKee and Ellen Nolte, citing OECD data, concluded that the UK had one of the worst rates of mortality amenable to healthcare among rich nations. If the NHS was to perform as well as the best-performing countries, thousands of lives could be saved each year. We make no apology for that.

Finally, Goldacre appears to have misunderstood the aims of our plans. We are not advocating reform for the sake of ideology. The changes we are proposing are designed to put patients first, improve health outcomes, empower clinicians so they can design services that meet the needs of their local communities, and put the NHS on a more sustainable footing so it is better able to meet the challenges of the 21st century.

Paul Burstow MP
Minister for Care Services

The government initially claimed that UK heart attack death rates were twice as bad as those in France. This was an overstatement: they are, but following recent interventions the gap is closing so rapidly that on current trends it will have disappeared entirely by 2012. In response, Burstow cites a 2008 paper by McKee and Nolte which he says 'concluded that the UK had one of the worst rates of mortality amenable to healthcare among rich nations'.

Burstow either misunderstands or misrepresents this very simple and brief paper. It is a study explicitly looking at time trends, not static figures, and it once again finds that comparing 2003 with 1998, the UK still had fairly high rates of avoidable mortality, but these were falling faster than in all but one of the other eighteen industrialised countries they examined. (Meanwhile, in the US, avoidable mortality improved at a disastrously slow pace, although they spent more money).

This is a paper showing the recent *success* of the NHS, and the fact that we are discussing such a massive improvement in avoidable mortality from Labour's first term in government is not my being partisan: this is the paper that was cited by the Tory Minister as evidence, bizarrely, of the NHS's recent failures.

Next, Burstow says I 'overlooked the impact assessment we published alongside the health and social care bill, where we present a thorough analysis of the evidence for and against our plans . . . studies show that GP fundholding and practice-based commissioning delivered shorter waits and fewer referrals to hospitals for patients'.

In its section on GP fundholding, this 'thorough analysis' ignores the four peer-reviewed academic papers I described last week, which sadly found no evidence of an overall benefit from GP fundholding. It makes a series of five assertions about outcomes, though not one of these is referenced to any single paper or study, anywhere at all.

I contacted the Department of Health, which ferreted out the sources especially for me. It turned out there was just one, a document from the King's Fund. It's not a peer-reviewed academic journal article, but the King's Fund is pretty good, in my view. If you read this document, once again, as with other reviews of the literature, it finds that the results of GP fundholding were mixed: some things got better, some things got worse.

So, the Minister has cherry-picked only the good findings,

from only one report, while ignoring the peer-reviewed literature. Most crucially, he cherry-picks findings he likes while
explicitly claiming that he is fairly citing the totality of the
evidence from a thorough analysis. By using this approach, I
can produce good evidence that I have a magical two-headed
coin, if I just disregard all the throws where it comes out tails.

Here is what politicians apparently cannot understand,
repeatedly, over and again: it's fine to make policy based on
ideology, whim, faith, principles, and all the other things we're
used to. It's also fine for evidence to be mixed. And it's absolutely fine if your reforms aren't supported by existing evidence:
just don't pretend that they are.

Politicians Can Divine Which Policy Works Best by Using Their Special Magic Politician Beam

Guardian, 22 May 2010

We have accidentally elected a coalition. This week all good citizens are poring over the 'Programme For Government', and there
is much to be pleased with. Labour wasn't all about unbridled
credit and fun public sector spending sprees: they kept all our
emails, kept records of the websites we visited, used 'anti-terrorism'
legislation on people who plainly weren't terrorists, and so on.

But most interesting are the noises now being made by the
coalition on crime and evidence. 'We will conduct a full review
of sentencing policy,' they say, 'to ensure that it is effective in
deterring crime, protecting the public, punishing offenders and
cutting reoffending. In particular, we will ensure that sentencing
for drug use helps offenders come off drugs.'

These are grand promises. Compulsory addiction rehabilitation with 'Drug Testing and Treatment Orders' was introduced precisely ten years ago – as an alternative to custodial sentences or simple probation – for those who have committed drug-related crimes. Their implementation without adequate analysis is a graphic example of our failure to run simple trials of social policy.

Any judge making a decision on a criminal's sentence is in the exact same position as a doctor making a decision on a patient's treatment: both are choosing an intervention for an individual in front of them, with the intention of producing a particular set of positive outcomes (reduced crime, say, and reduced drug use); both get through a large number of individuals in a month; and in many important situations, neither yet knows which of the available interventions works best.

If you randomly assign a fairly large number of criminals, or patients, to one of two interventions, in situations where you don't know which intervention would be best, and measure how well they're doing a year or so later, you instantly discover which intervention is best. Add in the cost, and you know which is most cost-effective. The basic principles behind this idea are not new, and were first described in the Old Testament, Daniel 1:12.

Before being rolled out nationally in October 2000, DTTOs were extensively piloted in three cities by the Criminal Policy Research Unit of South Bank University, at considerable cost. What insights did this generate? There was no randomisation, and no 'control' group of identical criminals given traditional sentences for comparison. Because of that, the only new knowledge generated by these pilots was the revelation that it is possible to set up a DTTO service and run it in some buildings in some cities.

As it happens, when they did follow up the people who had passed through the service, they hadn't done particularly well. But this finding wasn't published until after the service had been rolled out. In any case, because there was no randomised comparison group, we have no idea how these participants would have turned out if they'd been given a traditional custodial sentence anyway, so there's no sense in giving those results even a moment's thought.

This is a tragedy, and not just because drug use is estimated – with the usual caveats about estimating more nebulous stuff – to be behind 85 per cent of shoplifting, 80 per cent of domestic burglaries, over half of all robberies, and so on. It is also a tragedy because it speaks to motives that will never go away.

We would need a very brave modern politician to say: 'Look, I want to introduce a new policy, but I honestly don't know if it will work'. We might need to be forgiving, ourselves, of someone who said this, and actively encourage them to try out their ideas on half of a group of people. We would need a political class that could react to deferred outcomes, with the results dripping out from new interventions over the course of years, perhaps long after the one initiating politician has moved on. Doing all this would revolutionise social policy, in far wider domains than just criminal sentencing. But for now, politicians – and we must share the blame – get further when they use rhetoric, and absolutes.

Pornography in Hospitals

Guardian, 25 September 2010

The *Sun*, of all people, is angry about pornography: 'The hard-up NHS is blowing taxpayers' cash on PORN for sperm donors, a report reveals today.' The *Telegraph* immediately followed suit:

'Some clinics provide pornography for men masturbating in clinic rooms to produce sperm for IVF with their partners.'

These articles were inspired by a report titled 'Who said pornography was acceptable in the workplace', produced by a right-wing thinktank called 2020health. The author, former Conservative parliamentary candidate Julia Manning, says pornography in this clinical setting is: a violation of the NHS constitution; a case of manipulation by the sex industry; strips women of their human status; an encouragement of 'adultery of the mind'; a danger to men, as it introduces addictive material into their treatment (which 'beggars belief'); and an abuse of taxpayers' money.

The average spend on these magazines was £21.32 per health trust per year, with each clinic treating a large number of couples. For context, private clinics charge around £6,000 for a couple to have three cycles of IVF.

But the moral case may still stand: is the pornography necessary? Farmers, animal breeders and vets all have wide-ranging experience of getting viable sperm from male animals under artificial circumstances. As a result, they have examined this exact same question, in detail.

Hemsworth and Galloway showed in 1979 that sperm count in the ejaculate of a domestic boar (an actual boar, that's not a euphemism for men) was significantly increased by allowing a 'false mount', or observation of another boar's semen collection. We shouldn't overstate this evidence: another study found that the effect seems not to be present in rams. But in 1984 Mader and colleagues studied twelve Hereford bulls, and found that watching another mating pair in action significantly increased frequency of ejaculation. In the same year Price and colleagues found semen collection from male dairy goats was faster with a 'stimulus female', which was present but unmountable.

This can hardly be a surprise. As long ago as 1955, Kerruish reported that insemination centres for cows did not provide 'adequate sexual stimulation' prior to semen collection: his regimen of intensive sexual stimulation resulted in a 'marked improvement in sexual behaviour' and – crucially for our question – an increase in the conception rate.

But it gets more interesting. There is already evidence from animal research that males increase the amount of sperm in their ejaculate when there is more competition around. In 2005, Kilgallon and Simmons conducted an experiment to see whether human males viewing 'images depicting sperm competition' also had a higher percentage of motile sperm in their ejaculates.

This wasn't a perfect study: they compared ejaculate in fifty-two heterosexual men, looking at pornography with two men and one woman, against pornography with three women. I think it would have been better to compare images of one man and one woman, against two men and one woman, but there you go. They found that men viewing the 'two men one woman' pornography had a higher percentage of motile sperm. On a related note, Zbinden and colleagues found that male stickleback fish ejaculate more sperm after being shown a big rival than a small one.

But finally – and firmly on the question at hand – Yamamoto and colleagues in 2000 studied nineteen men masturbating into a jar, alone in a room, with or without 'sexually stimulating videotaped visual images'. Sperm volume, total sperm count, sperm motility and percentage of morphologically normal sperm were all higher when the men had pornography. Meanwhile, some men find it impossible to ejaculate on the day it's most needed for IVF, and sperm can only be retrieved by epididymal aspiration, or a needle inserted into the testicle. This is a seriously suboptimal outcome, with a small risk of unpleasant medical complications.

I'm not saying porn is brilliant. I absolutely agree that the objectification of women's bodies is a bad thing, and I don't particularly want to see porn lying around at work, although by the very nature of hospitals, you can see all kinds of dreadful things if you open the wrong door at the wrong time.

All I'm saying is: when there is a reasonable evidence base to show that pornography helps people overcome what they regard as a deeply painful problem (like 'being childless'); when they're going through the very strange and unpleasant experience of masturbating alone in a clinic room, with everyone outside knowing what they're doing, and quite possibly some kind of queue; then however unpleasant you might find the intervention, research showing that pornography works, matters.

The Power of Ideas

The Atheist's Guide to Christmas, 2009

I don't mean to fill your Christmas with Aids and diarrhoea, but there is something awe-inspiring about the power of ideas alone to do great good, and great evil. Diarrhoea will be our happy ending. Aids will not.

There are the cheap shots. Africa is filled with miracle-cure peddlers: the Gambian president, Yahya Jammeh, claims he can personally cure HIV, Aids and asthma using magic and charms. The South African government fell for a cure built around nothing more than industrial solvent.

It's all too easy to feel smug, and to forget that we have our own cultural idiosyncrasies. There's compelling evidence, after all, that needle-exchange programmes reduce the spread of HIV, but the strategy has been rejected, time and again, in favour of 'Just say no.'

And then there is the Church. In May 2009, as I write this, the Congolese Bishops' Conference have triumphantly announced that they 'say no to condoms!' This idiocy goes to the heart of the Catholic faith. In March, on his flight to Cameroon, Pope Benedict XVI explained that condoms worsen the Aids problem, and he has been supported, in the past year alone, by Cardinal George Pell of Sydney, Australia, and Cardinal Cormac Murphy O'Connor, the Archbishop of Westminster. 'It is quite ridiculous to go on about Aids in Africa and condoms, and the Catholic Church,' says O'Connor. 'I talk to priests who say, "My diocese is flooded with condoms and there is more Aids because of them." '

Some have been imaginative in promoting their message. In 2007, Archbishop Francisco Chimoio of Mozambique announced that European condom manufacturers were deliberately infecting

Casanova testing condoms

condoms with HIV to spread Aids in Africa. It is estimated that one in six people in Mozambique is HIV positive. Cardinal Alfonso López Trujillo of Colombia famously claimed that the HIV virus can pass through tiny holes in the rubber of condoms (air molecules are smaller than the HIV virus: blow a condom up, as a home experiment, to test if Trujillo's claim is true). 'The condom is a cork,' said Bishop Demetrio Fernandez of Spain, 'and not always effective.' In 2005 Bishop Elio Sgreccia, president of the Pontifical Academy for Life, explained that scientific research has never proven condoms 'immunise against infection'. He's right, I suppose. All this explains why the Pope has proclaimed: 'The most effective presence on the front in the battle against HIV/Aids is in fact the Catholic Church and her institutions.'

Meanwhile, development charities funded by US Christian

groups refuse to engage with birth control, and any suggestion of abortion – even in countries where being in control of your own fertility could mean the difference between success and failure in life – is met with a cold, pious stare. These moral principles are so deeply entrenched that under George W. Bush, the US Presidential Emergency Plan for Aids Relief insisted that every recipient of international aid money must sign a declaration expressly promising not to have any involvement with sex workers, even though they are a key vector for HIV.

Equally, there are heartbreaking tales of Westerners with a whiff of science going off to the developing world. Matthias Rath is a German vitamin-pill salesman who moved into South Africa five years ago, taking out full-page adverts in national newspapers: 'The answer to the Aids epidemic is here,' he announced. The answer, of course, was a vitamin pill. He explained that antiretroviral medications were a conspiracy by the pharmaceutical industry to kill patients and make money. 'Why should South Africans continue to be poisoned?' he asked.

And he had taken his ideas to the right place. In South Africa alone, 300,000 people die every year from the virus; that's one every two minutes. There are 1.2 million Aids orphans, and more than half of all pregnant women are HIV positive. And South Africa was headed by President Thabo Mbeki, an 'Aids dissident', as they prefer to be known. In the most crucial period of the Aids epidemic, the South African government variously claimed that HIV was not the cause of Aids, and that antiretroviral medications were not an effective treatment. It refused to roll out effective antiretroviral medication, it refused to accept gifts of money to give out ARV treatment, and it refused gifts of the pills themselves.

Mbeki's Health Minister would appear on television to talk up the risks of antiretroviral medication and talk down its benefits, promoting garlic and sweet African potato as effective treatments for Aids. The South African government's stall at the 2006 World Aids Conference in Toronto was described by other delegates as 'the salad stall', because that's all it contained. It has been estimated that between 2000 and 2005, around 350,000 people with Aids died unnecessarily in South Africa as a result of these ideas. That's quite a death toll, for some ideas.

How does this happen? Perhaps Aids is just too big to think about clearly. Twenty-five million people have died of it so far, three million in the past year: these figures are so vast, so overwhelming, that it's hard to mount an appropriate emotional response.

Perhaps the undeniable crimes of the pharmaceutical industry make conspiracy theories about its effective products believable, lending them a kind of poetic truth. It does, after all, adhere to cruel and murderous pricing policies, and only this year, regulations were explicitly changed so that clinical trials conducted by American companies on people in the developing world are no longer subject to the same high ethical standards as those conducted on US citizens.

Pharmaceutical companies, of course, are not all bad, just as there are many good people in the Catholic Church (the overwhelming majority, I would imagine). And a cheap, single dose of the drug Nevirapine, we should remember, has been shown to reduce the risk of a pregnant mother passing on HIV to her baby by half.

But we are mistaken if we imagine that medicine moves forward through technology. In the past that was probably true: antibiotics, intensive care units, and all the tools and technologies of modern medicine are dazzling. But much of the power lies with simple ideas.

So, let me tell you about diarrhoea. What are our two biggest weapons against torrential watery stool? One is telling people to wash their hands: it's been demonstrated that this can halve the spread of diarrhoea, and so it could save a million lives a year. The other is even simpler, and even more powerful: telling people to rehydrate, using water with added sugar and salt. This is new, it has been carefully researched and refined, and despite being a simple idea which anyone can follow, it has caused deaths from diarrhoea to plummet, saving at least fifty million lives since its universal adoption in the 1980s.

We can go higher. A hundred million people died in the last century from smoking. Around a billion are expected to die this century (because the cigarette industry has been so successful in China). But equally, tens of millions of lives will be saved, because Richard Doll and his colleagues diligently collected and analysed data on the smoking habits and deaths of a few thousand doctors fifty years ago to pull out just one key fact: smoking kills.

With this idea, with handwashing, with rehydration fluid, and with the methods and principles that gave us these ideas, a small group of softly-spoken saints have saved more people than you would meet in a thousand lifetimes.

The royalties from *The Atheist's Guide to Christmas* will go to the Terrence Higgins Trust. I have seen the work they do up close, working in an HIV clinic in South London. THT offers practical support, but the most powerful work their staff do, to my mind, is in sharing information, destigmatising, informing, preventing infections and improving compliance with treatment regimes. Through the application of common sense, wit, compassion and evidence, they save lives. This is the future of medicine.

And it's eight teaspoons of sugar and one of salt in a litre of water, if you ever need to know.

Merry Xmas.

'Exams Are Getting Easier'

Guardian, 21 August 2010

Pass rates are at 98 per cent. A quarter of grades are an A or higher. This week, every newspaper in the country was filled with people asserting that exams are definitely getting easier, and other people asserting that exams are definitely not getting easier. The question is always simple: how do you know?

Firstly, the idea of kids getting cleverer is not ludicrous. 'The Flynn Effect' is a term coined to describe the gradual improvement in IQ scores. This has been an important problem for IQ researchers, since IQ tests are peer referenced: that is, your performance is compared against everyone else, and the scores are rejigged so that the average IQ is always 100. Because of the trend to higher scores, year on year, you have to be careful not to use older tests on current populations, or their scores come out spuriously high, by the standards of the weaker average population of the past. Regardless of what you think about IQ tests, the tasks in them are at least relatively consistent. That said, there's also some evidence that the Flynn effect has slowed in developed countries recently.

But ideally, we want research that addresses exams directly. One approach would be to measure current kids' performance on the exams of the past. This is what the Royal Society of Chemistry did in its report 'The Five Decade Challenge' in 2008, running the project as a competition for sixteen-year-olds, which netted them 1,300 self-selecting higher-ability kids. They sat tests taken from the numerical and analytical components of O-level and GCSE exams over the past half-century, and it was found that performance against each decade rose over time: the average score for the 1960s questions was 15 per cent, rising to 35 per cent for the current exams (though with a giant leap around the introduction of GCSEs, after which the score remained fairly stable).

There are often many possible explanations for a finding. These results could mean that exams have got easier, but it's also possible that syllabuses have changed, so modern kids are less prepared for older-style questions. When the researchers looked at specific questions, they found that some things had been removed from the GCSE syllabus – because they'd moved up to A-level – but that's drifting unwelcomely towards anecdote.

Another approach would be to compare performance on a consistent test, over the years, against performance on A-levels. Robert Coe at Durham University produced a study of just this for the Office of National Statistics in 2007. Every year since 1988 a few thousand children have been given the Test of Developed Abilities, a consistent test (with a blip in 2002) of general maths and verbal reasoning skills. The scores saw a modest decline over the 1990s, and have been fairly flat for the past decade. But the clever thing is what the researchers did next: they worked out the A-level scores for children, accounting for their TDA scores, and found that children with the same TDA score were getting higher and higher exam results. From 1988 to 2006, for the same TDA score, A-level results rose by an average of two grades in each subject.

It could be that exams are easier. It could be that teaching and learning have improved, or that teaching has become more exam-focused, so kids at the same TDA level do better in A-levels: this is hard to measure. It could be that TDA scores are as irrelevant as shoe size, so the finding is spurious.

Alternatively, it could be that exams are different: they might be easier, say, with respect to verbal reasoning and maths, but harder with respect to something else. This, again, is hard to quantify. If the content and goals of exams change, then that poses difficulties for measuring their consistency over time, and it might be something to declare loudly (or to consult employers and the public about, since they seem concerned).

Our last study thinks more clearly along those lines: some people do have clear goals from education, and they can measure students against this yardstick, over time. 'Measuring the Mathematics Problem' is a report done for the Engineering Council and other august bodies in 2000, analysing data from sixty university maths, physics and engineering departments which gave diagnostic tests on basic maths skills to their new undergraduates each year. These were tests on things that mattered to university educators, and if something educational matters to them, we might think it matters overall. They found strong evidence of a steady decline in scores on their own tests, over the preceding decade, among students accepted onto degree courses where they would need good maths skills.

Sadly they didn't control for A-level grade, so we can't be sure how far they were comparing like with like, but there are various plausible explanations for their finding. Maybe maths syllabuses changed, and were less useful for maths and engineering degrees. Maybe the cleverest kids are doing English these days, or becoming lawyers instead. Or maybe exams got easier.

If you know of more research, I'd be interested to see it, but the main thing that strikes me here is the paucity of work in the field. There's a man called Rupert Sheldrake who believes that pets are psychic: they know when their owners are coming home, that sort of thing. Obviously we disagree on a lot, but we chat, and are friendly, and once when we were talking he came out with an excellent suggestion: maybe 1 per cent – or even 0.01 per cent – of the total UK research budget could be given to the public, so that they could decide what their research obsessions were. Maybe most of this money would get spent on psychic pets, or research into which vegetables cure cancer, but since we're all clearly preoccupied with the idea, I'd like to think that some of it, possibly, might get spent on good-quality, robust research to find out whether exams are getting easier.

Over There! An Eight-Mile-High Distraction Made of Posh Chocolate!

Guardian, 1 August 2009

This week the Food Standards Agency published two review papers showing that organic food is no better than normal food, in terms of composition, or health benefits. The Soil Association's response has been swift, and it has been given prominent, blanket right of reply throughout the media. That is testament to the lobbying power of a £2 billion industry, and the cultural values of journalists. I don't care about organic food, but I am interested in bad arguments. The Soil Association has made three.

Firstly, it says that the important issue with organic food is not the personal health benefit (there is none to be found in this data), but rather the benefit to the environment. This strategy – 'Don't talk about that, talk about this' – is a popular one, but it is cheap: we can talk separately about the environmental issues with organic food, but right now we are talking about the health effects.

Secondly, it says that the health benefits of organic food are related to pesticides, and cannot be measured by the evidence that has been identified and summarised in the FSA paper. This, once more, is gamesmanship. There was no evidence of health benefits to individuals. Possibly the Soil Association is proposing that there are health benefits which somehow cannot be measured: it is hard to disentangle the health benefits of eating organic food from other beneficial features of people's lives, for example, when you measure their health in a thirty-year study. In this case it is expressing a position of faith, not evidence. Or it is proposing that there are health benefits which could be measured, but have not been measured yet. In that case, again, this is faith rather than evidence, but it could start recruiting researchers now, using some small portion of its industry's £2 billion revenue to investigate these beliefs with fair tests.

Lastly, like many vitamin-pill peddlers and pharmaceutical companies before it, the Soil Association seeks to undermine the public's understanding of what a 'systematic review' is. It says that the report has deliberately excluded evidence to produce the answer that organic food is no better. The accusation is one of cherry-picking, and it is hard to see how it can be valid in the kind of studies that have been published here. These are 'systematic reviews': before you begin collecting papers to include, you specify how you will search for evidence, what databases you will use, what types of studies you will use, how you will grade the quality of the evidence (to see if it was a 'fair test'), and so on.

What does the Soil Association think these systematic reviews have ignored? As an example, from its press release, the industry body is 'disappointed that the FSA failed to include the results of a major European Union-funded study involving thirty-one research and university institutes and the publication, so far, of more than one hundred scientific papers, at a cost of 18 million Euros, which ended in April this year'. It gave the link to www. qlif.org.

On this website, you will find the QLIF list of 120 papers. Almost all are irrelevant. The first fourteen are on 'consumer expectations and attitudes', which are correctly not included in a systematic review of the evidence on food composition and health. Then there are twenty-two on 'effects of production methods'. Here you might expect to find more relevant research, but no.

The first paper ('The effect of medium term feeding with organic, low input and conventional diet on selected immune parameters in rat'), while interesting, will plainly not be relevant to a systematic review on nutrient content. The same is true of the next paper, 'Salmonella Infection Level in Danish

Indoor and Outdoor Pig Production Systems measured by Antibodies in Meat Juice and Faecal Shedding on-farm and at Slaughter': you might love its results, you might hate them; either way, it doesn't matter. This paper is simply not relevant to a review on nutrient content.

What's more, the overwhelming majority of these studies are unpublished conference papers, and some are just brief descriptions of the fact that somebody made an oral presentation at a meeting. The systematic review correctly looked only at good-quality data published in peer-reviewed academic journals, and with good reason: we know that conference papers are unreliable sources of information, that they often change between conference and publication, and that they are often never published at all.

As so often, this is about transparency, which is ultimately the only source of authority in science: we want the methods and results of scientific research to be formally presented, and accessible by all, so that we can see what was done, and what they found. If a government report on anything relies substantially on unpublished and inaccessible research, then we are correctly concerned. In fact, just two weeks ago this column discussed how the key piece of evidence presented by the Home Office to justify retaining DNA from innocent people who have only ever been arrested and then released was an incompetently presented piece of unpublished, incomplete research.

Systematic reviews give you transparency, because from the very outset everyone is honest and open about what kinds of studies they will and won't include. When organisations like the Soil Association greet the publication of a systematic review with accusations of cherry-picking, they're not just wrong, they're undermining the public's understanding of one of the most important and simple new ideas from the past three decades of science.

That annoys me, so let me share a prejudice. Ultimately, when people talk about the health benefits of organic food, they're not really talking about the health benefits of organic food. Wealthy society's big love for organic farming is about something very different: a bundle of legitimate concerns about unchecked capitalism in our food supply, battery farming, corruptible regulators, and reckless destruction of the environment, where the producers' costs do not reflect the true full costs of their activities to society, to name just a few. Every one of these problems deserves our full, individual attention.

But magic spells will not work. We cannot eradicate deceit from the pharmaceutical industry by buying homeopathic sugar pills; and we cannot solve the problems of unchecked capitalism in industrial food production by giving money to the £2 billion organic food industry in exchange for the occasional posh carrot.

As Far as I Understand
Thinktanks . . .

Guardian, 7 June 2008

There has been a frightening decline in the quality of maths in reports complaining about the frightening decline in the quality of maths in Britain. 'The Value of Mathematics', by thinktank Reform, has received large quantities of flattering media coverage this week from *The Times*, the *Telegraph*, and even scored a second puff in the *Guardian* from Professor Marcus du Sautoy, Oxford's Professor of the Public Understanding of Science. Their argument is simple: there is less maths around; people think it's cool to be bad at sums; we suffer economically; these are bad things.

Here is a key, early, factual claim from Reform's report: 'About 40 per cent of mathematics graduates enter financial services.'

This, they say, is a good thing. Do we believe the number? The report references it to the front page of a website called prospects. ac.uk, which is a pretty big website. Chasing through the pages there, you will find 'What Do Graduates Do?', and then the maths page. There were 4,070 maths graduates in their sampling frame for the year 2006. Only 2,010 of those, however, are in UK employment (1.5 per cent are working abroad, and the rest are studying for a higher degree, or a teaching qualification, or are unemployed, or unavailable for employment, and so on).

Of those 2,010 – not 4,070 – 37.9 per cent are working as 'Business and Financial Professionals and Associate Professionals'. So if we use maths: $2,010 \times 0.379 = 761.79$, and that divided by 4,070 gives us 0.1871, but let's round up like the angry maths profs did. About 20 per cent of maths graduates enter financial services. Not 40 per cent. For a group of people complaining about the substitution of woolly modern notions like 'relevance' and 'applied maths' in place of high-end mathematical techniques, these maths profs aren't very good at arithmetic.

They're also not very good at basic applied maths. For example, they argue that if we simply had more people around with maths knowledge, then there would automatically be more (and more lucrative) jobs requiring that knowledge, which our new maths graduates would instantly take up. I'm not an economist, but I'm not sure labour markets work like that: there are lots of men in the north of England who are very good at mining, and nobody in a hurry to open new pits.

Unfortunately, in any case, even this aspect of the report seems to be marred by simple errors in applied arithmetic. The thinktank is worried that the loss of A-level mathematicians has resulted in lost earnings for the economy. If the number of maths candidates had remained constant, they say, there would have been an additional 430,700 over the period 1989 to 2007. In the adjacent table they say 430,031, but that's the least of our worries. They go on to reason like this: 'Each of these students would have earned an additional £3,080 per year due to the market premium on A-level mathematics, equating to £136,000 over their lifetime. The total gain to the economy over the period would have been over £9 billion.'

We'll put aside the fact that the BBC said 'A maths A-level puts on average an extra £10,000 a year on a salary [not £3,080], says Reform,' because I can't get £9 billion for that period with those numbers (I get £12 billion, assuming a linear decline). Even if I could, Reform are making assumptions that are hard to blindly accept: in particular, will the extra earnings for people with maths A-level really still hold, if more people have maths A-level? We could go on.

I'm happy to agree that maths is economically useful, that some people might want to avoid difficult school subjects, and that humanities graduates who think maths is uncool are bores. What I would like is someone who can be bothered to sit down and reinforce my prejudices without perpetrating crass errors of over-interpretation and getting the basic arithmetic wrong. I've never fully seen the point of them, but until now, I assumed that's what thinktanks and Professors of the Public Understanding of Science are there for.

Meaningful Debates Need Clear Information

Guardian, 27 October 2007

Where do all those numbers in the newspapers come from? The Commons Committee on Science and Technology is taking evidence on 'scientific developments relating to the Abortion Act 1967'.

Scientific and medical expert bodies giving evidence say that survival in births below the twenty-fourth week of pregnancy has not significantly improved since the 1990s, when it was only 10–20 per cent. But one expert, a Professor of Neonatal Medicine, says survival at twenty-two and twenty-three weeks has improved. In fact, he says survival rates in this group can be phenomenally high: 42 per cent of children born at twenty-three weeks at some top specialist centres. He has been quoted widely: in the *Independent* and the *Telegraph*, on Channel 4 and *Newsnight*, by Tory MPs, and so on. The figure has a life of its own.

In the media, you get one expert saying one thing, and another saying something else. Who do you believe? The devil is in the detail. One option is to examine the messenger. John Wyatt is a member of the Christian Medical Fellowship. He didn't declare that when he went to give evidence. You don't have to. He did declare it when asked.

Prof Wyatt has relevant research experience, but there were half a dozen other medics without any relevant background who submitted evidence (or their view of it) to the committee who, when asked if they had anything to declare, did mention membership of Christian or evangelical groups with an established position on abortion. I don't care for an argument that rests on competing ideologies, so let's look at Prof Wyatt's evidence: because it has been hugely reported, and it goes against the evidence from a huge study called Epicure.

Epicure contains all of the data for every premature birth in England over the course of a year: one snapshot was taken in 1995, and one in 2006. Overall, it shows a modest improvement in survival for births at twenty-four weeks, but no significant improvement in the 10–20 per cent rate for births at twenty-two and twenty-three weeks.

For the next bit, you need to remember one simple piece of primary school maths. In the figure 3/20, 3 is the numerator and 20 is the denominator. If you have three survivors for every twenty births, then 15 per cent survive. For Epicure, the numerator is survival to discharge from hospital, and the denominator is all births where there is a sign of life, carefully defined.

There are two ways you could get a higher survival percentage. One would be a genuine increase in the number of babies surviving, an increase in the numerator: eight out of twenty live births survive, 40 per cent. But you could also see an increase in the survival percentage by changing the denominator. Let's say, instead of counting as your denominator 'all births where there is a sign of life in the delivery room', you counted 'all babies admitted to neonatal intensive care'. Now, that's a different kettle of fish altogether. To be admitted to neonatal ICU, the doctors have to think you've got a chance. Often you have to be transferred from another hospital in an ambulance, and for that you really do have to be more well. Therefore, if your denominator is 'neonatal ICU admissions', your survival rates will be higher, but you are not comparing like with like. That may partly explain Prof Wyatt's figure for a very high survival rate in twenty-three-week babies. But it's not clear.

First, in his written evidence he said that the data was from a 'prospectively defined' study (where the researchers say in advance what they plan to collect). Then he was asked in the committee, when giving his oral evidence: 'What was the denominator for that? Was that . . . 42 per cent survival at twenty-three weeks of all babies showing signs of life in the delivery room, or was it a proportion of those admitted to neonatal intensive care directly or by transfer?' Prof Wyatt replied: 'The denominator was all babies born alive in the labour ward in the hospital at UCL [University College London].' This later turned out not to be true.

Then he was asked to send the reference for the claim. He did so. It was merely an abstract for an academic conference presentation three years ago. It did not contain the figures he was quoting. He then said he had done the raw figures on a spreadsheet, especially for the committee – bespoke, if you will – and sent them in. They are entered into the record as a memo, on 18 October 2007. They show new, different, but broadly comparable figures: 50 per cent survive at twenty-two weeks, then down to 46 per cent at twenty-three weeks, then up to 82 per cent at twenty-four weeks, then down again to 77 per cent at twenty-five weeks. (That bouncing around is because the raw numbers are so small that there is a lot of random noise.)

And the denominator? Prof Wyatt is clear: 'I have provided the numbers and percentage of infants born alive at University College London Hospitals who survived to one year of age.' The committee asked for clarification of this. Finally, on 23 October, another memo arrived from Prof Wyatt, entered into the record where all can read it. For the widely quoted 42 per cent survival rate at twenty-three weeks, Prof Wyatt admitted that the denominator was, in fact, all babies admitted to the neonatal intensive care unit. But in his new special analysis, giving this new '46 per cent survive at twenty-three weeks' figure, the figures in the previous paragraph, he claimed the denominator was 'all live births'. Has he undone a prospectively designed study, and retrospectively redesigned it? Or is this now a completely different source of data to the original reference?

I cannot blame Prof Wyatt for this, but his figure has taken on a life of its own. There may have been yet another mistake here, about the denominator. I don't know. I'm quite prepared to believe that UCL may have unusually good results. But science is about clarity and transparency, especially for public policy. You need to be very clear on things like: what do you define as a 'live birth', how do you decide on what gestational age was, and so on. Even if this data stands up eventually, right now it is non-peer-reviewed, unpublished, utterly chaotic, changeable, personal communication of data, from 1996 to

2000, with no clear source, and no information about how it was collected or analysed. That would be fine if it hadn't suddenly become central to the debate on abortion.

Minority Retort

Guardian, 3 November 2007

Parliamentary select committees are among the few places where you can see politicians sitting down and doing the kind of thing you'd actually want them to do, like thinking carefully about policy. This week the Science and Technology Committee delivered its report on scientific developments relating to the Abortion Act. This is, entirely for free, a fantastic piece of popular science writing on epidemiology. In particular, it's a masterclass on spotting dodgy statistics, which is exactly what the committee received from anti-abortion activists.

Here is one example, on the question of abortion and breast cancer risk, in which a parliamentary document explains the importance of choosing the correct control group.

Dr Richards told us that 'if you compare women who keep their pregnancy with those who have an induced abortion, those who have an induced abortion are more likely to get breast cancer later on'. This is the comparative group that Dr Brind favours and the result is expected, since carrying a first pregnancy to birth is protective against breast cancer. However, if you look at the rates of cancer between women who have had an abortion and those who have not had children, the effect disappears.

This is the bread and butter of science; a thing to behold. They give similarly rigorous and transparent treatment to the foetal pain people, the neonatal survival figures, and more.

Two Conservative MPs on the committee who favour tighter restrictions on abortion were unhappy with this report. They have issued their own minority report, published as an appendix to the main one. Does this differ in approach, or moral values? No, it differs in something much more simple: the quality of the science, the selectivity of the quoting, and the quality of the referencing.

There isn't space to debunk it in this column (and, bafflingly, most aspects of it are already debunked by the main report it accompanies). But if you want one good illustration of their approach, then you might want to read the bit where they talk about, well, me:

> We were greatly concerned to read in the *Guardian* on 27 October an article clearly aimed at undermining the credibility of Professor John Wyatt which contained detailed information about Wyatt's evidence . . . which could only have been passed on to the journalist concerned by a member of the select committee. There should be an inquiry about how this information got into the public domain and as to whether such a personal attack represents a serious breach of parliamentary procedure.

My article did contain detailed information about Prof Wyatt's evidence. But I suspect that any inquiry set up to examine how I managed to obtain that information would finish its work well before the first set of tea and biscuits arrived, since all the facts came from the written and oral evidence, published openly and in full on the parliament.uk website during the select committee hearing. I downloaded the PDF, and then I read it. If there is a lesson for parliamentarians here, it is a simple one: we're watching you, and we're allowed to.

Building Evidence into Education*

I think there is a huge prize waiting to be claimed by teachers. By collecting better evidence about what works best, and establishing a culture where this evidence is used as a matter of routine, we can improve outcomes for children, and increase professional independence.

This is not an unusual idea. Medicine has leapt forward with evidence-based practice, because it's only by conducting 'randomised trials' – fair tests, comparing one treatment against another – that we've been able to find out what works best. Outcomes for patients have improved as a result, through thousands of tiny steps forward. But these gains haven't been won simply by doing a few individual trials, on a few single topics, in a few hospitals here and there. A change of culture was also required, with more education about evidence for medics, and whole new systems to run trials as a matter of routine, to identify questions that matter to practitioners, to gather evidence on what works best, and then, crucially, to get it read, understood, and put into practice.

I want to persuade you that this revolution could – and should – happen in education. There are many differences between

---

* Writing is just a hobby, alongside seeing patients, doing research, and – increasingly – putting these ideas into practice through campaigning and lobbying. In 2011 I co-authored a Cabinet Office White Paper explaining how randomised controlled trials can be used to improve government policy (this is the subject of my next book). And in 2012 I did an Independent External Review for the Department for Education, looking at what could be done to improve the use of evidence and data in schools. This was a dry internal report, but I was also asked to write something to explain what these changes might look like, aimed specifically at teachers, which is reproduced here. If you like it, and want to share it, there's a PDF on the DfE website. I feel fairly hopeful, but these are long, slow cultural shifts. In June 2014 DfE advertised a public tender to assess progress towards the goals I set out in the internal report; so we shall see.

medicine and teaching, but they also have a lot in common. Both involve craft and personal expertise, learnt over years of experience. Both work best when we learn from the experiences of others, and what worked best for them. Every child is different, of course, and every patient is different too; but we are all similar enough that research can help find out which interventions will work best overall, and which strategies should be tried first, second or third, to help everyone achieve the best outcome.

Before we get that far, though, there is a caveat: I'm a doctor. I know that outsiders often try to tell teachers what they should do, and I'm aware this often ends badly. Because of that, there are two things we should be clear on.

Firstly, evidence-based practice isn't about telling teachers what to do – in fact, quite the opposite. This is about empowering teachers, and setting a profession free from governments, ministers and civil servants who are often overly keen on sending out edicts, insisting that their new idea is the best in town. Nobody in government would tell a doctor what to prescribe, but we all expect doctors to be able to make informed decisions about which treatment is best, using the best currently available evidence. I think teachers could one day be in the same position.

Secondly, doctors didn't invent evidence-based medicine. In fact, quite the opposite is true: just a few decades ago, best medical practice was driven by things like eminence, charisma and personal experience. We needed the help of statisticians, epidemiologists, information librarians and experts in trial design to move forwards. Many doctors – especially the most senior ones – fought hard against this, regarding 'evidence-based medicine' as a challenge to their authority.

In retrospect, we've seen that these doctors were wrong. The opportunity to make informed decisions about what works best, using good-quality evidence, represents a truer form of professional independence than any senior figure barking out his opinion. A coherent set of systems for evidence-based practice listens to people on the front line, to find out where the uncertainties are, and decide which ideas are worth testing. Lastly, crucially, individual judgement isn't undermined by evidence: if anything, informed judgement is back in the foreground, and hugely improved.

This is the opportunity that I think teachers might want to take up. Because some of these ideas might be new to some readers, I'll describe the basics of a randomised trial, but after that, I'll describe the systems and structures that exist to support evidence-based practice, which are in many ways more important. There is no need for a world where everyone is suddenly an expert on research, running trials in their classroom tomorrow: what matters is that most people understand the ideas, that we remove the barriers to 'fair tests' of what works, and that evidence can be used to improve outcomes.

How randomised trials work

Where they are feasible, randomised trials are generally the most reliable tool we have for finding out which of two interventions works best. We simply take a group of children, or schools (or patients, or people); we split them into two groups at random; we give one intervention to one group, and the other intervention to the other group; then we measure how each group is doing, to see if one intervention achieved its supposed outcome any better.

This is how medicines are tested, and in most circumstances it would be regarded as dangerous for anyone to use a treatment today, without ensuring that it had been shown to work well in a randomised trial. Trials are not only used in medicine, however, and it is common to find them being used in fields as diverse as web design, retail, government, and development work around the world.

For example, there was a long-standing debate about which of two competing models of 'microfinance' schemes was best at getting people out of poverty in India, whilst ensuring that the money was paid back, so it could be re-used in other villages: a randomised trial compared the two models, and established which was best.

At the top of the page at Wikipedia, when it is having a funding drive, you can see the smiling face of Jimmy Wales, the founder, on a fundraising advert. He's a fairly shy person, and didn't want his face to be on these banners. But Wikipedia ran a randomised trial, assigning visitors to different adverts: some saw an advert with a child from the developing world ('She could have access to all of human knowledge if you donate . . .'); some saw an attractive young intern; some saw Jimmy Wales. The adverts with Wales got more clicks and more donations than the rest, so they were used universally.

It's easy to imagine that there are ways around the inconvenience of randomly assigning people, or schools, to one intervention or another: surely, you might think, we could just look at the people who are already getting one intervention, or another, and simply monitor their outcomes to find out which is the best. But this approach suffers from a serious problem. If you don't randomise, and just observe what's happening in classrooms already, then the people getting different interventions might be very different from each other, in ways that are hard to measure.

For example, when you look across the country, children who are taught to read in one particularly strict and specific way at school may perform better on a reading test at age seven, but that doesn't necessarily mean that the strict, specific reading method was responsible for their better performance. It may just be that schools with more affluent children, or fewer social problems, are more able to get away with using this (imaginary) strict reading method, and their pupils were always going to perform better on reading tests at age seven.

This is also a problem when you are rolling out a new policy, and hoping to find out whether it works better than what's already in place. It is tempting to look at results before and after a new intervention is rolled out, but this can be very misleading, as other factors may have changed at the same time. For example, if you have a 'back to work' scheme that is supposed to get people on benefits back into employment, it might get implemented across the country at a time when the economy is picking up anyway, so more people will be finding jobs, and you might be misled into believing that it was your 'back to work' scheme that did the job (at best, you'll be tangled up in some very complex and arbitrary mathematical modelling, trying to discount for the effects of the economy picking up).

Sometimes people hope that running a pilot is a way around this, but this is also a mistake. Pilots are very informative about the practicalities of whether your new intervention can be implemented, but they can be very misleading on the benefits or harms, because the centres that participate in pilots are often different from the centres that don't. For example, job centres participating in a 'back to work' pilot might be less busy, or have more highly motivated staff: their clients were always going to do better, so a pilot in those centres will make the new jobs scheme look better than it really is. Similarly, running a pilot of a fashionable new educational intervention in schools that are already performing well might make the new idea look fantastic, when in reality the good results have nothing to do with the new intervention.

This is why randomised trials are the best way to find out how well a new intervention works: they ensure that the pupils or schools getting a new intervention are the same as the pupils and schools still getting the old one, because they are all randomly selected from the same pool.

At around this point, most people start to become nervous: surely it's wrong, for example, to decide what kind of education a child gets, simply at random? This cuts to the core of why we do trials, and why we gather evidence on what works best.

Myths about randomised trials

While there are some situations where trials aren't appropriate – and where we need to be cautious in interpreting the results – there are also several myths about trials. These myths are sometimes used to prevent trials being done, which slows down progress, and creates harm, by preventing us from finding out what works best. Some people even claim that trials are undesirable, and even completely impossible, in schools: this is a peculiarly local idea, and there have been huge numbers of trials in education in other countries, such as the US. However, the specific myths are worth discussing.

Firstly, people sometimes worry that it is unethical to randomly assign children to one educational intervention or another. Often this is driven by an implicit belief that a new or expensive intervention is always necessarily better. When people believe this, they also worry that it's wrong to deprive people of the new intervention. It's important to be clear, before we get to the detail, that a trial doesn't necessarily involve depriving people of anything, since we can often run a trial where people are randomly assigned to receive the new intervention now, or after a six-month wait. But there is a more important reason why trials are ethically acceptable: in reality, before we do a trial, we generally have no idea which of two interventions is best. Furthermore, new things that many people believe in can sometimes turn out, in reality, to be very harmful.

Medicine is littered with examples of this, and it is a frightening reality. For many years, it was common to treat everyone who had a serious head injury with steroids. This made perfect sense on paper: head injuries cause the brain to swell up, which can cause important structures to be crushed inside our rigid skulls; but steroids reduce swelling (this is why you have steroid injections for a swollen knee), so they should improve survival. Nobody ran a trial on this for many years. In fact, it was widely argued that randomising unconscious patients in A&E to have steroids or not would be unethical and unfair, so trials were actively blocked. When a trial was finally conducted, it turned out that steroids actually increased the chances of dying after a head injury. The new intervention, that made perfect sense on paper, that everyone believed in, was killing people: not in large enough numbers to be immediately obvious, but when the trial was finally done, an extra two people died out of every hundred given steroids.

There are similar cases from the world of education. The 'Scared Straight' programme also made sense on paper: young children were taken into prisons and shown the consequences of a life of crime, in the hope that they would be more law-abiding in their own lives. Following the children who participated in this programme into adult life, it seemed they were less likely to commit crimes, when compared with other children. But here, researchers were caught out by the same problem discussed above: the schools – and so the children – who went on the Scared Straight course were different from the children who didn't. When a randomised trial was finally done, where this error could be accounted for, we found out that the Scared Straight programme – rolled out at great expense, with great enthusiasm, good intentions and huge optimism – was actively harmful, making children more likely to go to prison in later life.

So we must always be cautious about assuming that things which are new, or expensive, are necessarily always better. But this is just one special case of a broader issue: we should always be clear when we are uncertain about which intervention is best. Right now, there are huge numbers of different interventions used throughout the country – different strategies to reduce absenteeism, or teach arithmetic, or reduce teenage pregnancies, or any number of other things – where there is no evidence to say which of the currently used methods is best. There is arbitrary variation, across the country, across a town, in what strategies and methods are used, and nobody worries that there is an ethical problem with this.

Randomisation, in a trial, adds one simple extra chink to this existing variation: we need a group of schools, teachers, pupils or parents who are able to honestly say: 'We don't know which of these two strategies is best, so we don't mind which we use. We want to find out which is best, and we know it won't harm us.'

This is a good example of how gathering good evidence requires a culture shift, extending beyond a few individual randomised trials. It requires everyone involved in education to recognise when it's time to honestly say 'We don't know what's best here.' This isn't a counsel of despair: in medicine, and in teaching, we know that most of what we do does some good (if we're not better than nothing, then we're all in big trouble!). The real challenge is in identifying what works the best, because when people are deprived of the best, they are harmed too. But this is also a reminder of how inappropriate certainty can be a barrier to progress, especially when there are charismatic people who claim they know what's best, even without good evidence.

Medicine suffered hugely with this problem, and as late as the 1970s there were notorious confrontations between people who thought it was important to run fair tests, and 'experts', who were angry at the thought of their expertise being challenged and their favourite practices being tested. Archie Cochrane was one of the pioneers of evidence-based medicine, and in his autobiography he describes many battles he had with senior doctors, in glorious detail. In 1971, Cochrane was concerned that coronary care units in hospitals might be no better than home care, which was the standard care for a heart attack at the time (we should remember that this was the early days of managing heart attacks, and the results from this study wouldn't be applicable today). In fact, he was worried that hospital care might involve a lot of risky procedures that could even, conceivably, make outcomes worse for patients overall.

Because of this, Cochrane tried to set up a randomised trial comparing home care against hospital care, despite great resistance from the cardiologists. In fact, the doctors running the new specialist units were so vicious about the very notion of running a trial that when one was finally set up, and the first results were collected, Cochrane decided to play a practical joke. These initial results showed that patients in coronary care units did worse than patients sent home; but Cochrane switched the numbers around, to make it look as if patients on CCUs did better. He showed the cardiologists these results, which reinforced their belief that it was wrong of Cochrane even to dare to try running a randomised trial of whether their specialist units were helpful. The room erupted: 'They were vociferous in their abuse: "Archie," they said "we always thought you were unethical. You must stop this trial at once." . . . I let them have their say for some time, then apologised and gave them the true results, challenging them to say as vehemently, that coronary care units should be stopped immediately. There was dead silence and I felt rather sick because they were, after all, my medical colleagues.'

Similar confrontations are reported in many fields, when people try subjecting ideas and practices to fair tests, in randomised trials. But being open and clear about the need for research – when there is no good evidence to help us choose between interventions – is also important, because it helps make sure that research is done on relevant questions, meeting the needs of teachers, pupils and parents. When everyone involved in teaching knows a little about how research is done – and what previous research has found – then we can all have a better idea of what questions need to be asked next.

But before we get on to how this can happen, we should

first finish the myths about trials. From now on, these are all cases where people overstate the benefits of trials.

For example, sometimes people think that trials can answer everything, or that they are the only form of evidence. This isn't true, and different methods are useful for answering different questions. Randomised trials are very good at showing that something works; they're not always so helpful for under-standing *why* it worked (although there are often clues when we can see that an intervention worked well in children with certain characteristics, but not so well in others). 'Qualitative' research – such as asking people open questions about their experiences – can help give a better understanding of how and why things worked, or failed, on the ground. This kind of research can also be useful for generating new questions about what works best, to be answered with trials. But qualitative research is very bad for finding out whether an intervention has worked. Sometimes researchers who lack the skills needed to conduct or even understand trials can feel threatened, and campaign hard against them, much like the experts in Archie Cochrane's story. I think this is a mistake. The trick is to ensure that the right method is used to answer the right questions.

A related issue involves choosing the right outcome to measure. Sometimes people say that trials are impossible, because we can't capture the intangible benefits that come from education, like making someone a well-rounded member of society. It's true that this outcome can be hard to measure, although that is an argument against any kind of measurement of attainment, and against any kind of quantitative research, not just trials. It's also, I think, a little far-fetched: there are lots of things we try to improve that are easy to measure, like attendance rates, teenage pregnancy, amount of exercise, perfor-mance on specific academic or performance tests, and so on.

However, we should return to the exaggerated claims some-times made in favour of trials, and the need to be a critical consumer of evidence. A further common mistake is to assume that, once an intervention has been shown to be effective in a single trial, then it definitely works, and we should use it everywhere. Again, this isn't necessarily true. Firstly, all trials need to be run properly: if there are flaws in a trial's design, then it stops being a fair test of the treatments. But more importantly, we need to think carefully about whether the people in a trial of an intervention are the same as the people we are thinking of using the intervention on.

The Family Nurse Partnership is a programme that is well funded and popular around the world. It was first shown to be effective in a randomised trial in 1977. The trial participants were white mothers in a semi-rural setting in upstate New York, and people worried at the time that the positive results might have been exceptional, and occurred simply because the specific programme of social support that was offered had suited this population unusually well. In 1988, to check that the findings really were applicable to other settings, the same programme was assessed using a randomised trial in African-American mothers in inner-city Memphis, and was again found to be effective. In 1994, a third trial was conducted in a large population of Hispanic, African-American and Caucasian mothers from Denver. After this trial also showed a benefit, people in the US were fairly certain that the programme worked, with fewer childhood injuries, increased maternal employment, improved 'school readiness', and more.

Now the Family Nurse Partnership programme is being brought to Britain, but the people who originally designed the intervention have insisted that a randomised trial should be run here, to see if it really is effective in the very different setting of the UK. They have specifically stated that they expect to see less dramatic benefits here, because the basic level of support for young families in the UK is much better than that in the US: this means that the difference between people getting the

FNP programme and people getting the normal level of help from society will be much smaller.

This is just one example of why we need to be thoughtful about whether the results of a trial in one population really are applicable to our own patients or pupils. It's also an illustration of why we need to make trials part of the everyday routine, so that we can replicate them in different settings, instead of blindly assuming we can use results from other countries (or even other schools, if they have radically different populations). It doesn't mean, however, that we can never trust the results of a trial. This is just another example of why it's useful to know more about how trials work, and to be a thoughtful consumer of evidence.

Lastly, people sometimes worry that trials are expensive and complicated. This isn't necessarily true, and it's important to be clear what the costs of a trial are being compared against. For example, if the choice is between running a trial, and simply charging ahead, implementing an idea that hasn't been shown to work – one that might be ineffective, wasteful, or even harmful – then it's clearly worth investing some time and effort in assessing its true impact. If the alternative is doing an 'observational' study, which has all the shortcomings described above, then the analysis can be so expensive and complex – not to mention unreliable – that it would have been easier to randomise participants to one intervention or the other in the first place.

But the mechanics and administrative processes for running a trial can also be kept to a minimum with thoughtful design, for example by measuring outcomes using routine classroom data that was being collected anyway, rather than running a special set of tests. More than anything, though, for trials to be run efficiently, they need to be part of the culture of teaching.

Making evidence part of everyday life
I'm struck by how much enthusiasm there is for trials and evidence-based practice in some parts of teaching; but I'm also

struck that much of this enthusiasm dies out before it gets to do good, because the basic structures needed to support evidence-based practice are lacking. As a result, a small number of trials are done, but these exist as isolated islands, without enough bridges joining the people and strands of work together. This is nobody's fault: creating an 'information architecture' out of thin air is a big job, and it might take decades. The benefits, though, are potentially huge. Some individual randomised trials from the UK have produced informative results, for example, but these results are then poorly communicated, so they don't inform and change practice as well as they might.

Because of this, I've sketched out the basics of what education would need, as a sector, to embrace evidence-based practice in a serious way. The aim – which I hope everyone would share – is to get more research done, involving as many teachers as possible; and to get the results of good-quality research disseminated and put into practice. It's worth being clear, though, that this is a first sketch, and a call to arms. I hope that others will pull it apart and add to it. But I also hope that people will be able to act on it, because structures like these in medicine help capture the best value from the good work – and hard work – that is done all around the country.

Firstly – and most simply – it's clear that we need better systems for disseminating the findings of research to teachers on the ground. While individual studies are written up in very technical documents, in obscure academic journals, these are rarely read by teachers. And rightly so: most doctors rarely bother to read technical academic journals either. The *British Medical Journal* has brief summaries of important new research from around the world; and there is a thriving market of people offering accessible summary information on new 'what works' research to doctors, nurses and other healthcare professionals. The US government has spent vast sums of money on two similar websites for teachers: 'Doing What Works', and the 'What Works Clearing

House'. These are large, with good-quality resources, and they are written to be relevant to teachers' needs, rather than dry academic games. While there are some similar resources in the UK, these are often short-lived, and on a smaller scale.

For these kinds of resources to be useful at all, they then need to land with teachers who know the basics of 'how we know' what works. While much teacher training has reflected the results of research, this evidence has often been presented as a completed canon of answers. It's much rarer to find all young teachers being taught the basics of how different types of research are done, and the strengths and weaknesses of each approach on different types of question (although some individual teachers have taught themselves on this topic, to a very high level). Learning the basics of how research works is important, not because every teacher should be a researcher, but because it allows teachers to be critical consumers of the new research findings that will come out during the many decades of their career. It also means that some of the barriers to research that arise from myths and misunderstandings can be overcome. In an ideal world, teachers would be taught this in basic teacher training, and it would be reinforced in Continuing Professional Development, alongside summaries of research.

In some parts of the world, it is impossible to rise up the career ladder of teaching without understanding how research can improve practice, and publishing articles in teaching journals. Teachers in Shanghai and Singapore participate in regular 'Journal Clubs', where they discuss a new piece of research, and its strengths and weaknesses, before considering whether they would apply its findings in their own practice. If the answer is no, they share the shortcomings in the study design that they've identified, and then describe any better research that they think should be done on the same question.

This is an important quirk: understanding how research is done also enables teachers to generate new research questions.

This, in turn, ensures that the research which gets done addresses the needs of everyday teachers. In medicine, any doctor can feed up a research suggestion to NIHR (the National Institute for Health Research), and there are organisations that maintain lists of what we don't yet know, fed by clinicians who've had to make decisions, without good-quality evidence to guide them. But there are also less tangible ways that this feedback can take place.

Familiarity with the basics of how research works also helps teachers to get involved in research, and to see through the dangerous myths about trials being actively undesirable, or even 'impossible', in education. Here, there is a striking difference with medicine. Many teachers pour their heart and soul into research projects which are supposed to find out whether something worked; but in reality the projects often turn out to be too small, being run by one person in isolation, in only one classroom, and lack the expert support necessary to ensure a robust design. Very few doctors would try to run a quantitative research project alone in their own single practice, without expert support from a statistician, and without help from someone experienced in research design.

In fact, most doctors participate in research by playing a small role in a larger research project which is coordinated, for example, through a research network. Many GPs are happy to help out on a research: they recruit participants from among their patients; they deliver whichever of two commonly used treatments has been randomly assigned to their patient; and they share medical information for follow-up data. But they get involved by putting their name down with the Primary Care Research Network covering their area. Researchers interested in running a randomised trial in GP patients then go to the Research Network, and find GPs to work with.

This system represents a kind of 'dating service' for practitioners and researchers. Creating similar networks in education would help join up the enthusiasm that many teachers have for research that improves practice with researchers, who can sometimes struggle to find schools willing to participate in good-quality research. This kind of two-way exchange between researchers and teachers also helps the teacher-researchers of the future to learn more about the nuts and bolts of running a trial; and it helps to keep researchers out of their ivory towers, focusing more on what matters most to teachers.

In the background, for academics, there is much more to be said on details. We need, I think, academic funders who listen to teachers, and focus on commissioning research that helps us learn what works best to improve outcomes. We need academics with quantitative research skills from outside traditional academic education departments – economists, demographers, and more – to come in and share their skills more often, in a multidisciplinary fashion. We need more expert collaboration with Clinical Trials Units, to ensure that common pitfalls in randomised trial design are avoided; we may also need – eventually – Education Trials Units, helping to support good-quality research throughout the country.

But just as this issue stretches way beyond a few individual research projects, it also goes way beyond anything that one single player can achieve. We are describing the creation of a whole ecosystem from nothing. Whether or not it happens depends on individual teachers, researchers, heads, politicians, pupils, parents, and more. It will take mischievous leaders, unafraid to question orthodoxies by producing good-quality evidence; and it will need to land with a community that – at the very least – doesn't misunderstand evidence-based practice, or reject randomised trials out of hand.

If this all sounds like a lot of work, then it should do: it will take a long time. But the gains are huge, and not just in terms of better evidence, and better outcomes for pupils. Right now, there is a wave of enthusiasm for good-quality evidence, passing through all corners of government. This is the time to act. Teachers have the opportunity, I believe, to become an evidence-based profession, in just one generation: embedding research into everyday practice; making informed decisions independently; and fighting off the odd spectacle of governments telling teachers how to teach, because teachers can use the good-quality evidence that they have helped to create, to make their own informed judgements.

There is also a roadmap. While evidence-based medicine seems like an obvious idea today – and we would be horrified to hear of doctors using treatments without gathering and using evidence on which works best – in reality these battles were only won in very recent decades. Many eminent doctors fought viciously, as recently as the 1970s, against the very idea of evidence-based medicine, seeing it as a challenge to their expertise. The case for change was made by optimistic young practitioners like Archie Cochrane, who saw that good evidence on what works best was worth fighting for.

Now we recognise that being a good doctor, or teacher, or manager, isn't about robotically following the numerical output of randomised trials; nor is it about ignoring the evidence, and following your hunches and personal experiences instead. We do best by using the right combination of skills to get the best job done.

DRUGS

A Rock of Crack as Big as the Ritz*

Guardian, 21 February 2009

In a week where our dear *Daily Mail* ran with the headline 'How Using Facebook Could Raise your Risk of Cancer', I will exercise some self-control, and write about drugs instead.

'Seven hundred British troops seized four Taliban narcotics factories containing £50m of drugs,' said the *Guardian* on Wednesday. 'Troops recovered more than 400kg of raw opium in one drug factory and nearly 800kg of heroin in another.' That is good. In the *Telegraph*, British forces had seized '£50 million of heroin and killed at least twenty Taliban fighters in a daring raid that dealt a significant blow to the insurgents in Afghanistan'. Everyone carried the good news. 'John Hutton, Defence Secretary, said the seizure of £50m of narcotics would "starve the Taliban of funding preventing the proliferation of drugs and terror in the UK".'

Well.

First up, almost every paper – the people we pay to précis facts for mass consumption – got both the quantities and the substances wrong. From the MoD press release (a romping read), three batches of opium were captured, but no heroin:

---

* I chose titles on my blog spontaneously when I posted the pieces. 'A Rock of Crack as Big as the Ritz' is the title of a short story by Will Self.

'over 60kg of wet opium', 'over 400kg of raw opium' and 'the largest find of opium on the operation, nearly 800kg'.

So the army captured 1,260kg of opium. Opium is not heroin; it takes about 10kg of opium to make 1kg of heroin. They also found some chemicals and vats. The opium was enough to make roughly 130kg of heroin.

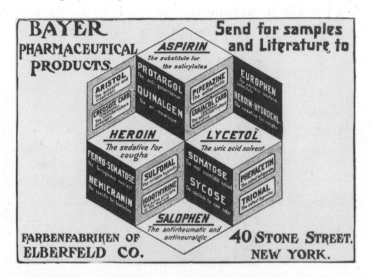

How much was this haul worth to the Taliban, and exactly how much of a blow will it strike? Heroin is not very valuable in itself, because opium is easy to grow, and you can turn it into heroin over the course of three simple steps using some school science-class chemicals in your kitchen (or a barn in rural Afghanistan). Heroin becomes expensive because it is illegal, so people must take risks to produce and distribute it, and as a result, they want money.

The 'farm gate' price of 1kg of opium in Afghanistan is $100 at best (we'll use dollars, since the best figures are from the UN Drugs Control Programme 2008 world report). So the 1,260kg of opium captured on this raid in Afghanistan is worth somewhere near $126,000 (not £50 million).

Even if it had been converted to heroin – it wasn't – the money doesn't get much better. The price for 1kg of heroin in Afghanistan is not much higher than the price for the 10kg of opium you need to make it. That's because heroin was invented over a hundred years ago, and making it, as I said, is cheap and easy. We could be generous and say that heroin is worth $2,000 per kilo in Afghanistan: this would make the army's (potential) 130kg of heroin worth about $250,000.

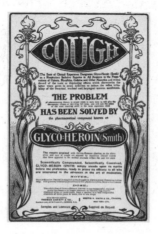

That's still not £50 million. So where did this enormous number come from? Perhaps everyone was trying to calculate it by using the wholesale price in the UK, assuming that the Taliban ran the entire operation from 'farm gate' to 'warehouse in Essex'. This is a stretch of our generosity, but we can still run the numbers: the wholesale price of heroin in the UK has fallen dramatically over the past two decades, from $54,000 per kilo in 1990 to $28,000 today. That would make our 130kg of (potential) heroin worth $3.6 million.

We're still nowhere near £50 million. So: maybe the army thinks that every sweaty kid with missing teeth in King's Cross selling £10 bags is actually a Taliban agent, passing profits on – in full, no cream off the top – to Taliban HQ, several thousand miles away in Aghanistan. Even then, UK heroin is $71 per gram at retail prices (down from $157 a gram in 1990), so the value of our 130kg is $9 million. We can be generous, and say our street heroin is only 30 per cent pure (it's usually much better): in this case, finally, the haul is worth $30 million on the streets, or £20 million, at absolute best. To do this, we had to assume that every penny of the street-level UK retail price, at the smallest unit of sale, went straight to the Taliban, and that's not £50 million.

But the most important thing about figures – once you've actually got them right – is to put them in their appropriate context. Even if we were generous, would 130kg less heroin make any difference to the UK market? No. We consume tons and tons of heroin every year, and the heroin from Afghanistan, in any case, is going all around the world, not just into the UK.

More importantly, would this seizure make much difference to the Taliban, whichever figure you use: $126,000, or $3.6 million, or $30 million, or £50 million? Again, that seems unlikely. There are 157,000 hectares (100 metres squared) of opium fields in Afghanistan, producing 7,700 *tons* (not kilos) of opium, netting farmers throughout the country about $730 million, and that's real money in their pocket, not made-up UK street prices on the diluted gram. The export value of opium, morphine and heroin at border prices in neighbouring countries for Afghan traffickers was worth $3.4 billion last year.

Just to remind you: John Hutton is the Defence Secretary, and he said that the seizure of £50 million of narcotics would 'starve the Taliban of funding preventing the proliferation of drugs and terror in the UK'. That frightens me, because I trust the Defence Secretary to know what's going on in a *war*, and you didn't even need to do the maths on his figure: this seizure was a tiny drop of theatre in a very, very big ocean.

The Least Surrogate Outcome

Guardian, 5 April 2008

There's this vague idea – which has been going around for the past few centuries – that statistics is difficult. But in reality the maths is often the least of your problems: the tricky bit comes way before the number crunching, when you are deciding what to measure, how to measure it, and what those measurements mean.

The government's new drugs strategy has been published, with outcomes that will be measured to see if it works or not. However you cut the cake, we should be clear: measuring drug-related death is difficult. You could look at death certificates to see what's listed, but they're often filled out by junior doctors, and aren't very informative or reliable. You also need to decide where to draw the causal cut-off. Does HIV count as a drug-related death, if you got it from a needle full of heroin? Or from sex work to fund the drugs?

How about if it kills you ten years after you become abstinent, or you die from chronic, grumbling hepatitis C from a needle? Or chronic, seeping, pus-ridden abscesses bulging deep in your groin from years of injecting your femoral veins?

And that's before we get to crack-frenzy violence and drug driving. What if there was no toxicology done? What if there was, but they didn't test for the drug the person took? What if the coroner finds some drugs in the blood, but doesn't think they were related to the death? Are they consistent in making that call?

The new government drugs strategy solves this tricky problem by simply not measuring drug-related deaths as an outcome any more. It was a key indicator in our major strategy document ten years ago, but you won't see death mentioned once in 'Drugs: Protecting Families and Communities Action Plan 2008–2011.

You also won't see death in 'Public Service Agreement Delivery Agreement 25', which includes measured outcomes such as the number of users in treatment and the rate of drug-related offending. A lot of drug users die. Death, even if you don't like drug users, is important.

But beyond the disputes over how you collect these figures, there is the interpretation and analysis; and the greatest irony is that the government may have dropped drug-related deaths two weeks ago, simply because it misunderstood that the figures are actually looking quite good.

Overall, drug-related deaths show no great improvement over the years. But what if older people – over thirty-five, say, users from the great injecting epidemic of the 1980s – were dying at a greater rate, while young people, the target of great effort, are dying at a slower rate? That's what a recent analysis from the biostatistics unit in Cambridge shows: they presented their findings just two weeks ago, in the same week, by odd coincidence, that the government announced its deathless drug strategy.

Sometimes people can be so stupid that they don't even know when they've done well.

Heroin on Prescription

This is an essay I wrote in praise of heroin prescription, as an undergraduate in medicine, which won the 'Roger Hole Essay Prize in Medical Scepticism' at my medical school in 1998. The £250 prize was useful, but winning it also served up an early lesson on the flaws in science and health reporting, and its obses-

sion with 'authority'. A friend in a thinktank gave my number to somebody working on campaigning journalist Nick Davies' heroin addiction documentary in 2000. They called, said they'd heard I was a doctor and had researched the issue, and asked if I would do an interview.

I was happy to help, but explained that I'd only qualified as a doctor two weeks previously, that I wasn't an 'authority' on anything, that I wouldn't even be fully registered with the GMC for another year, and that I looked about twelve. I chatted through what I knew on the subject, and casually offered to email them an essay I had written as a medical student. I did that, and never heard back.

A few months later I switched on the telly halfway through a TV show about heroin on prescription. It didn't occur to me that this might be related to the brief phone call I'd had, but suddenly the screen went black, with a dramatic pause and – if I remember right – a deep bass synthesiser tone in the background. Then, from nowhere, quotes from my medical student essay appeared, in big sombre letters, filling the screen in white on black, ascribed to 'Dr Ben Goldacre', as if I was a grand medical authority. I was sat on a mattress on the floor, eating toast in my underpants, aged twenty-five, in a shared flat with no hot water or heating. I've never been so embarrassed. I prayed that nobody I worked with had seen it. It's one reason why I still feel uneasy being described as 'Dr Ben Goldacre'. Anyway, here's the essay. Forgive the pompous writing – I was twenty-three.

Methadone and Heroin:
An Exercise in Medical Scepticism
by Ben Goldacre, 1998

I have often fantasised about living through an age when science was truly adversarial: to have seen Darwin at the Royal Society, or Galileo recant. But the lie of the land, the structure of our scientific territory, and our modes of warfare across it, have become domesticated and tame. Although there may be differences of opinion, we each tend to tinker at the expanding edges of our understanding, and truly mutually exclusive explanatory frameworks for reproducible phenomena rarely co-exist for long.

If we want to see real friction, some other factor must come to bear on the essentially healthy structure of the mainstream scientific community: a funding issue, for example, might influence the general trajectory of research, but for the most part temporarily; we may be transiently confounded by partisan research from a given drug company, but only in whatever microcosm of physiology their drug acts and, albeit slightly behind schedule, we can be sure that the truth will out.

But we want the big prize: we want wholesale irrationality, we want to see the axial skeleton of our concepts truly and chronically deformed, and only politics can muster contorting forces of such magnitude. I intend to show how this influence has perverted rationality in one area, our medical treatment of those who are addicted to heroin, to such an extent that our theory and practice is now so polluted as to be scientifically untenable.

Until recently, it was common practice in Britain to prescribe heroin to heroin addicts. This apparently paradoxical practice was well founded and successful, as we shall see below. However, since the late 1960s, addicts have mostly been prescribed oral methadone, a long-acting opiate agonist with a less euphoriant effect, as a heroin substitute.

I shall demonstrate that the maintenance of addicts on methadone is less effective at reducing the use of heroin, and the harm that goes along with heroin use, than the prescription of heroin itself. I shall also show that methadone is a more dangerous drug than heroin, and causes more deaths than even adulterated street heroin.

Ultimately, the case I shall make is that heroin prescription

228

is more effective, by all reasonable outcome indicators, than methadone; and that the reasons for its unpopularity have little to do with evidence of best practice, and much to do with our emotional and moral attitudes towards those who are addicted to drugs. To begin, we must consider the history of opiate addiction and its treatment, in order to understand how and why politics intervened, and how we arrived at the state we are in today.

A Brief History of Opiate Addiction and Rehabilitation

Heroin, or diamorphine, was first marketed by Bayer in 1898, after being developed as a cough suppressant by the same team who introduced aspirin. Although the use of psychoactive drugs (and specifically opiates) began to be considered as a medical problem in the late nineteenth century, they were legally available until well into the twentieth, and opiates could be purchased from pharmacies with a minimum of formalities up until 1920.

At this time, the principal medical concern over drug use was the risk of accidental poisoning through the non-medical pursuit of pleasure, and the prevalence of drug addiction (frequently caused by chronic medical use) was so low that its social impact was negligible. The 1920 Dangerous Drugs Act confined the availability of opiates to prescription only and, over the next few years, penalties for offenders against the possession laws were increased, a reflection of similar developments abroad.

However, where Britain departed from the rest of the world was with the Rolleston Committee report from the Department of Health in 1926. This emphasised that persistent drug use, in line with newly emerging medical and social theory, should be seen as a disease: 'as a manifestation of a morbid state, and not as a mere form of vicious indulgence'.

By pursuing this line, Rolleston arrived almost accidentally at the sympathetic modern-day conception of the drug abuser, over half a century before Hartnoll et al. (1980) found evidence of serious childhood disturbance in his patients at a drug dependency clinic in University College Hospital. In many ways Rolleston was the first proponent of the guiding philosophy of most modern drug work, 'harm reduction', which I shall later consider in detail.

The progressive attitude to drug use institutionalised in this report established the framework of public policy for the next five decades, and following 1926 the 'British System' prosecuted dealers and dilettantes, but permitted medical prescription of heroin to addicts after 'every effort' had been made for the 'cure of the addiction', but when the drugs could not be fully withdrawn without 'severe distress or even risk of life' or 'experience showed that a certain minimum dose of the drug was necessary for the patients to lead useful and relatively normal lives . . . capable of work'. This twin policy of 'policing and prescribing' effectively contained the heroin problem (which ran at below a hundred notified heroin addicts) for the next four decades.

With the sixties came an atmosphere of moral panic at the scale of a well-publicised increase in drug use. Although the drugs in question were mostly cannabis and amphetamines, not heroin, attitudes to drug use and regulation were reappraised: amphetamines and LSD were brought under tight statutory control, and the government began to fear that with a rising demand for drugs, the licit opiate supply system might start supplying the illicit market.

From 1959 to 1964 the number of addicts notified to the Home Office increased from sixty-eight to 342, and it was noted that an unusually large proportion of these new addicts were of non-therapeutic origin, that is, an unusually large proportion of new users had not come to addiction via chronic medical treatment for physical disease or injury.

In 1964, the government convened the Brain Committee, an

interdepartmental reincarnation of the Rolleston Committee, who found that 'the major source of supply had been excessive prescribing by a small group of doctors'. They recommended that the prescription of drugs to addicts should be restricted to specialist clinics, 'Drug Dependency Units' (DDUs), and although heroin for physical ailments could still be freely prescribed, laws were passed requiring that doctors who prescribed heroin for addicts should be specifically approved by the Home Secretary.

From the beginning of the seventies, there was a major sea change in the treatment of addicts. This was characterised by an emphasis on abstinence as the primary goal of treatment, and a refusal to prescribe heroin: instead, on condition of abstinence from all other drugs, and under 'treatment contracts', heroin addicts were prescribed a new drug, methadone, to be taken orally.

It has been proposed that the reluctance of doctors to prescribe heroin was probably, in a number of cases, due to the fact that most addicts were so keen to obtain this drug: this made doctors working in the field uneasy, and Glossop (1995) believes that prescribing a medicine which was less desirable for the client was more easily rationalised.

In the mid-1970s there was an upsurge in the illicit heroin market in London. The factors alleged to have contributed to this include: an upsurge in illicit demand following the change in DDU policy; the end of the Vietnam War, requiring the South-East Asian producers to find new markets after the GIs went home; wealthy Iranian exiles using heroin as a means of getting their capital out of the country after the downfall of the Shah; and political troubles in Afghanistan and Pakistan.

Over the course of the decade the market ceased to be run by amateurs, as professional criminals extended their interests from the cannabis market to heroin. Driven in part by a pyramid dealing network, where addicts at the bottom of the distribution chain had an interest in selling to fund their own use, heroin use expanded enormously throughout the next twenty years.

The Contemporary Heroin Problem

The Home Office were notified of 35,000 heroin addicts in 1994; due to incomplete reporting, and other obstacles in reaching the addict population, this is widely believed to represent between a third and a fifth of all addicts, thus putting the true figure at between 100,000 and 160,000.

Contemporary heroin addiction is no longer an issue between the individual and their metabolism. The nature of the drug, the scale of its use, and its position in modern society, all mean that addicts experience more diverse problems, and cause more diverse problems, than the heroin addict of the nineteenth century.

The generally poor health of chronic addicts is usually not a direct result of the opiates as such. Heroin is very addictive but does not in itself cause any serious illnesses, nor does it harm any organs or tissues: the indirect consequences are of course more serious. Pain sensations are suppressed, with the results that certain signals (for example, problems with teeth, infections, cold, heat and hunger) are not noticed. Because opiates also suppress feelings and emotions, the ability to enter into social relations with others is also seriously affected, so not only physical but social functioning worsens.

Another important issue is how the addict can maintain a supply of heroin. The enormous cost of heroin on the black market is met for the most part by acquisitive property crime. The economics of the illicit market are remarkable: at the farm gate in Pakistan, a kilogram of opium costs $90; when it has been converted to heroin it costs $3,000 in Pakistan; wholesale in the USA it costs $80,000; and its final retail price on the street (at the Drug Enforcement Agency's quoted average purity of 40 per cent) is $290,000 per kilogram.

On the streets of the UK, a gram of heroin costs between £50 and £120. The cost of acquisitive crime committed to pay for this heroin has been estimated at £1.5 billion per year. Addicts in the UK generally steal to fund their addiction: thus they risk likely impoverishment and imprisonment. One study showed that 80 per cent of addicts attending a DDU clinic had been convicted of at least one offence in the course of their drug-taking careers. More crucially for long-term outcome, since they often steal from family and friends, addicts risk social isolation.

Furthermore, since the drug is at such a premium, it will be used in the most efficient fashion possible, which is of course intravenous injection (intravenous use of alcohol under prohibition of alcohol in the USA has also been documented). Intravenous use of any drug carries its own dangers. A large proportion of the morbidity experienced amongst heroin addicts is due to wound infection, septicaemia and infective endocarditis, all due to asterile injection technique.

Infection is another major cause of morbidity and mortality in intravenous drug users internationally. Heroin addicts tend to lead chaotic lifestyles and have low self-esteem, both of which, along with expediency, contribute to a tendency to share needles with other users. Via this route they become infected with HIV, and hepatitis B and C.

Ten per cent of UK Aids cases in 1995 were related to the use of intravenous drugs, and it is suspected that the increase in HIV infection amongst non-intravenous drug using heterosexuals is being driven by contact with heterosexual intravenous drug users, and the World Health Organization estimates that 40 per cent of recent Aids cases internationally were caused by drug users sharing injecting equipment.

It was the spread of HIV through intravenous drug use that led to the reconsideration of heroin addiction treatment in the late 1980s, and was the birth of the new policy of 'harm reduction'. The HIV seropositivity rate amongst intravenous drug users in Edinburgh, where needle-exchange and maintenance programmes had been vigorously opposed, rose to over 50 per cent in the mid-1980s. By comparison, in Glasgow, where such facilities were available, less than fifty miles away, the level of seropositivity was less than 5 per cent.*

A policy of harm reduction tackles public health issues directly by seeking to reduce the personal and social costs of drug use. Abstinence is not regarded as a realistic short-term goal for most dependent users, and the principal ingredients of most programmes are syringe exchanges, educational and advisory services, and treatment and maintenance services (generally with methadone).

There is a hierarchy of achievable objectives, with non-users urged to abstain, and users advised to reduce doses, and to avoid the most potent drugs and riskier means of ingestion. Those who insist on injecting are offered advice on safer technique, and those who persist in sharing needles are even taught how to clean their equipment.

This policy has been vigorously opposed in some parts of the world, especially the USA, where drug-related mortality is almost twice that of the UK. Despite this, it has become the guiding principle behind UK drugs policy, along with the maintenance prescription of methadone.

Methadone

Methadone is an opioid receptor agonist with a half-life of approximately twenty-four hours, far longer than heroin. Drugs with longer half-lives tend to produce less acute withdrawal effects, a phenomenon which is utilised in the choice of anxiolytic drugs in psychiatric practice. Crucially, in comparison with heroin, methadone has a greatly reduced euphoric effect. The

---

* There might be other explanations; I was only a medical student – BG, 2014.

hope for methadone, therefore, is that it can contain the opiate cravings, on a once-daily oral dose, without providing so much of a 'high'.

The aim of methadone maintenance is to stabilise and then to 'cure' the opiate user. This breaks down into such objectives as: improving the health of drug users, by providing clean drugs in measured doses under medical supervision; reducing drug-related crime by providing users with free legal opiates, thus reducing their need to steal to fund illicit heroin; improving the social situation of drug users (family relationships, finances, employment, housing and so on); persuading users to reduce their daily dose and ultimately take steps towards abstinence. This is in many ways an updated version of Rolleston's rationale from 1926.

However, the policy of prescribing methadone may be criticised from many different angles, and to the best of my knowledge these criticisms have never been comprehensively considered in one article. Certainly there is no convenient meta-analysis of methadone programmes. I shall consider each criticism in detail, and later compare the use of methadone to the maintenance prescription of heroin, which still continues on a small scale in the UK, and has recently been reassessed in Switzerland and Australia.

Firstly, it is important to recognise that methadone is not a pleasant drug to take, causing nausea and vomiting, weight gain, profuse sweating, dysphoria and tooth decay. This is no major selling-point to a patient group clearly accustomed to making stringent aesthetic judgements about their drugs, and this, combined with the absence of the 'buzz' of heroin, means that the take-up rate amongst addicts is far lower than it ever was for heroin.

Hartnoll et al. (1980) found that only 29 per cent of those offered methadone in one DDU between 1972 and 1975 were still attending twelve months later. The reality of take-up rates for methadone prescription programmes amongst the general population of heroin addicts today is that only a small minority of addicts will attend methadone clinics, certainly less than 15 per cent, although specific statistics are hampered by the unknown quantity of the denominator, that is, the number of people in a population addicted to illicit drugs.

Treatment for drug dependency, to be successful and especially to have an impact at a community level, must have high take-up and retention rates amongst problem drug users who, unlike adults with right iliac fossa pain, may not spontaneously present themselves to healthcare professionals.

In order to be successful, therefore, a drug dependency unit must offer both treatment and the drug at a 'price' which the users are willing to pay: the prescriptions may be free, but the terms and conditions on which they are offered may act as a deterrent to some users, and the product offered (counselling services, advice, and possibly substitute drugs) must be appealing. Health economists have couched the problem in their own terminology: 'For treatment to have a high take-up rate, it must sell . . . and be seen to sell . . . a good product at low cost.'

Retention in treatment, firstly, is an area where the philosophy guiding the work of a clinic may have as much of an impact as the nature of the drug it is offering. In a controlled study in Australia, heroin addicts were assessed and randomised to two clinics, one oriented to long-term methadone maintenance, and the other oriented to time-limited treatment, aiming primarily at abstinence from all drugs, including methadone. Both groups were urine-tested for heroin, and use of heroin outside the clinic was higher in the abstinence-oriented clinic.

An observational study in a different country showed that addicts were more likely to discharge themselves earlier from methadone clinics where the clinic staff scored highly on an 'Abstinence Orientation Scale', measuring their commitment to abstinence-oriented policies on heroin addiction. Other studies have shown that external compulsion to attend clinics, for example by law courts, is also associated with poor retention.

Conversely, a high re-attendance rate has been demonstrated at 'user-friendly' clinics where needle exchange and clean drugs are available, with no uninvited counselling. Experience has taught that regular and enduring contact with treatment services is a necessary precondition for successful treatment of addicts.

Finally, studies of drug users who present to rehabilitation programmes have shown that they are often in a poorer state of health than other heroin addicts in the population (of equally long standing) who have not chosen to present, and this is taken by some commentators to mean that addicts will only present as a last resort. Thus methadone programmes are by no means a universally attractive option to the addict population, and addicts often use their drug of choice to supplement their prescription.

Use of heroin outside the confines of a drug-rehabilitation programme (whilst ostensibly attending it) is, of course, associated with all of the risks of everyday heroin addiction: increased risk of intravenous drug use leading to infection, increased acquisitive crime, poor family relations. More importantly, the chaotic nature of the drug use means that the chances of abstinence after a period of regulated drug use are reduced. Thus use of heroin outside the clinic may be considered one of the definitively poor outcome measures.

However, methadone is also a dangerous drug in its own right: astonishingly, use of methadone has a higher mortality even than the use of illicit heroin, although to what extent is uncertain. For example, in 1992, there were 101 deaths from methadone, and forty from heroin; similarly, from 1982 to 1991 there were 349 methadone deaths and 243 heroin deaths: this is despite the fact that there are far more users of heroin, at every stratum of use, by a factor of at least 3:2, than of methadone.

However, to quantify the mortality requires an accurate

denominator (the number of users for each drug), and this, as we have already discussed, can only be achieved indirectly for a covert and underground activity such as drug abuse. Estimates vary widely according to the denominator used, and authors are never so disingenuous as to claim pinpoint accuracy for their figures, but the most recent data to be analysed estimates the risk of methadone-related mortality at around four times that of heroin.

The dangers of methadone have long been recognised. Ghodse et al. (1985) analysed the patient records of notified addicts who died in the UK between 1967 and 1981, and found that among patients using heroin, three quarters of deaths were directly drug related, and 'most deaths in which a drug was implicated were due to medically prescribed drugs' (invariably methadone). A retrospective cohort study followed up 128 addicts who first presented in London in 1969, of whom twenty-eight had died, and reported similar findings.

Reasons for this high mortality have been ascribed to its long half-life: a large number of deaths occur in the first few days of treatment, and this may be due to the chronic accumulation of the methadone in the bodies of addicts with reduced liver function. Other reasons proposed include black-market consumption, which is harder to quantify, and the co-administration of heroin and methadone, for which there is less evidence, albeit that death certificates provide notoriously poor data.

Clearly there is a paucity of mortality data in the literature on methadone prescription. In 1994, a review of the methodology of drug treatment evaluation found that only four out of seventeen UK studies had used mortality as an outcome measure. To neglect this most 'ineffective' of outcomes, in studies of a drug which is prescribed to 17,000 British addicts, in whom it has a demonstrably higher mortality than the drug it is substituted for, seems extraordinary.

Finally, and perhaps most bizarrely, it is generally recognised that methadone is a more addictive drug than heroin, with a more arduous withdrawal process, and this fact is recognised both among the drug-using subculture and in the scientific literature.

Heroin on Prescription

The current situation is that very little heroin is prescribed in the UK: it was estimated that 117 addicts were prescribed heroin in 1992, while 17,000 were prescribed methadone. Maintenance prescription of heroin, the 'British System' until the 1960s, is the ultimate extension of harm reductionist philosophy. There are many deductive arguments to support it, but little modern experimental data, and many criticisms that are laid against it. I shall consider these extensively, before examining the few studies of contemporary heroin maintenance programmes which have recently been published.

The philosophy behind the prescription of heroin owes a lot to the findings of the Rolleston Committee in 1926, is similar to the thinking behind methadone prescription, and is essentially as follows: addiction itself is not something that is readily amenable to medical intervention, and as such opiates are prescribed to the addict for as long as they remain addicted, in order to keep them in a state of good health and leading as normal (and crime-free) a life as possible.

Addiction has been famously characterised by Vaillant (1991) as a chronic relapsing condition with a spontaneous remission rate of 5 per cent per annum regardless of external intervention. This apparently flippant description is supported by empirical data on long-term follow-up of addicts which show that no external agency expedites the ending of addiction, not even major life events.

With drug addiction, we are often choosing between problems, rather than solutions, and so heroin maintenance, which is only ever offered to patients who have failed with other modalities of treatment, could be considered the best of a bad lot. 'With readily available prescribed opiates, there is no need to commit acquisitive crime to buy drugs, to sell drugs to others to finance one's own use, and to risk one's own (and others') health, not to mention life, with adulterated drugs of unknown strength.' It is also likely to promote attendance at the clinic for intervention when deemed appropriate, and an important side effect is the denial to criminals of a lucrative source of income.

There are of course a number of criticisms of heroin prescription. The first is that it negates the deterrent effect of the criminal law. However, heroin addicts already resist the deterrent effects of arrest, imprisonment, beatings by gangsters, social isolation, and injury or death through adulterants and disease. It is hard to imagine any greater sanctions than these, and so for addicts of this nature the choice may not be between detoxification or prescribed heroin, but between heroin from the illicit market or heroin from a clinic.

The second criticism is the possibility that heroin prescription would increase drug use in the general population. However, there is good evidence that untreated addicts must indulge in low-level and aggressive marketing of heroin to provide themselves with a supply; that is, they push the drug in order to obtain it, thus promoting increased general consumption. It was partly the cessation of maintenance prescription in the 1960s that led to the arrival of an aggressive black market. Furthermore, it seems likely that the improved contact with family, friends and healthcare providers that comes with maintenance prescription improves the chances of a healthy productive lifestyle and ultimate abstention.

This criticism of increased general use is linked to the fear of leakage of prescribed heroin onto the black market. This problem is best addressed by the careful prescription of an exact dose by specialist prescribers, and the evidence from the one remaining Rolleston clinic in the UK, or rather from the local drugs squad (who undertook to examine all arrested addicts for evidence of drugs prescribed by the clinic), was that there was no leakage onto the black market.

A final criticism of heroin prescription is that it is expensive, costing up to ten times more per year, per addict, than methadone. Firstly it is important to recognise that the cost of any drug is not the sole factor in the running of a nationwide treatment and rehabilitation programme. Public expenditure on drug control was £500 million in 1995, and of this £60 million was spent on treatment and rehabilitation, while £350 million was spent on police and customs enforcement, deterrents and control (which prevents less than 15 per cent of drugs from arriving on the black market).

If we calculate that there were 20,000 recipients of £500 of methadone annually, that is £10 million from the treatment budget, which would be £100 million if heroin was prescribed to a similar number. It seems likely that take-up and retention rates in clinics would increase if heroin was prescribed. Thus it would seem that this is perhaps the most viable of all our criticisms, and local health authorities have criticised heroin prescription on grounds of cost.

However, it has been claimed that because only one company may distribute heroin for medicinal purposes in the UK (Evans Medical), a virtual monopoly has been created, to the point where heroin is overpriced by a factor of thirty. This monopoly position was addressed by the European Court of Justice in 1995, who ruled that the government would have to open up the market to competition, but as yet there are no plans to change the situation.

We must now consider the studies which have sought to compare heroin and methadone. Such data is extremely thin on the ground: there was one small randomised control trial in UCH from 1972–75; and one similar study in Switzerland in 1995; there is also poorly quantified data from the one

Rolleston clinic in the UK which closed in 1995, and one small case-control study from Northern Ireland.

Hartnoll et al. (1980) studied ninety-six addicts randomised to oral methadone (OM) or heroin maintenance (HM). After twelve months, 74 per cent of the HM group were still attending, but only 29 per cent of the OM group had maintained contact with the clinic. For both groups, illicit heroin use decreased, in the HM group from 74 to 21 mg/day, but in the OM group from 74 mg to 37 mg/day. There was no difference between the two groups in employment status, but over the course of the year 32 per cent of the OM group had spent some time in prison, whereas 19 per cent of the HM group did so.

Perneger et al. (1998) studied fifty-one patients randomised similarly to heroin or methadone for six months in Geneva in 1995. At follow-up, there was a significant difference in use of street heroin, with 48 per cent of the OM group using street heroin on a daily basis, as opposed to 4 per cent of the HM group ($p = 0.002$). There was also a significant difference in the amount of money spent on drugs between the two groups, with the OM group spending approximately ten times the amount of the HM group ($p = 0.039$). Associated with this, the HM group were less likely to be charged with theft ($p = 0.015$) and less likely to be charged with drug dealing or possession ($p = 0.067$ and $p = 0.008$ respectively); overall, 57 per cent of the OM group were charged with any offence over the course of the six-month trial, whereas 19 per cent of the HM group were charged ($p = 0.0004$).

There was also a significant difference in mental-health status, with six suicide attempts in the OM group, against one in the HM group ($p = 0.022$). Finally, health-related quality of life was measured with the SF-36 scale, which found a greater improvement for the HM group in mental health ($p = 0.025$) and social functioning ($p = 0.041$). There was no differential improvement in employment status, housing situation, or somatic health between the two groups over six months.

McCusker and Davies (1996) found similar results at a clinic in Northern Ireland over six months. The HM group manifested lower levels of psychopathology and showed greater retention in treatment, criminal activity was significantly more reduced, as was illicit heroin use, and although there were again no differences in physical health, the OM group was the only one to report the sharing of used injecting equipment.

Thus our three trials demonstrated many advantages to the prescription of heroin, and none to methadone.

Conclusions

Clearly there are stout grounds for scepticism concerning the validity of prescribing methadone in the treatment of heroin addiction. It is also clear that there is a paucity of research in this field, a failure which must surely be redressed.

There are certain things of which we can be certain. Heroin is the most attractive drug for heroin addicts, and however we might wish them to behave, they continue to use illicit sources of the drug even if a substitute is prescribed. Methadone is undoubtedly a dangerous drug, and one that retards entry of addicts into the treatment programmes offered; it is also a drug whose effects have not been comprehensively researched. Heroin maintenance may well prove to be the best option we have.

Drug addiction is not a phenomenon that lends itself generously to empirical investigation. Even the outcome indicators are a subject for debate, and a viable study of what many would see as the ultimate index of success, abstention, would require a trial lasting more than a decade.

However, quantifiable indices of health status, social functioning, criminal behaviour, total opiate consumption, needle-sharing and so on are all viable and uncontroversial outcome measures, and should be comprehensively investigated for methadone and heroin. Furthermore, no indications have been found that prescribing heroin would inflict harm of a kind that might make such trials unacceptable.

Perneger et al. (1998) have noted that although the Swiss trial was small, it was similar to the initial evaluations of methadone, such as the seminal paper by Dole and Nyswander (1965), which led to its widespread use in the treatment of drug addiction. It seems likely that a contributory factor was the medical profession's emotional and moral attitudes towards drug users.

However noble our intentions when we approach a clinical or social problem, we may often be confounded by extraneous factors and preconceptions, and fail in our objectivity. We share an obligation to submit all medical interventions to rigorous, continuous and objective reassessment. Drug addiction affects 100,000 people in Britain directly, and many more indirectly; it is responsible for an enormous drain on health-care resources, a large proportion of acquisitive crime, and the fastest-growing group of HIV infection. That we should apparently neglect our obligations in such an important field is astonishing.

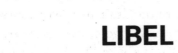

LIBEL

NMT Is Suing Dr Peter Wilmshurst. So How Trustworthy Is This Company? Let's Look at Its Website . . .

Guardian, 11 December 2010

You will hopefully remember – from the era before WikiLeaks – that US medical-device company NMT is suing NHS cardiologist Peter Wilmshurst over his comments about the conduct and results of the MIST trial, which sadly for NMT found no evidence that their device prevents migraine. The MIST trial was funded by NMT, and Wilmshurst was lead investigator until problems arose.

Wilmshurst has already paid £100,000 of his own money to defend himself, risking his house, and has spent every weekend and all his annual leave, unpaid, dealing with this, at great cost to his family. So what kind of a company is NMT Medical, which the British libel courts have allowed to hound one man for almost two years? And how trustworthy are its utterances?

Let's go to its website and find out. On the front page you will see positive quotes from three patients prominently displayed, on a rotating banner (reload the page to see the full collection), accompanied by smiling studio photographs.

At this point we should remember that the outcome of the MIST trial – the study in question here – really was negative. It set out to see if the device permanently prevents migraine. One hundred and forty-seven patients with migraine took part. Seventy-four had the NMT STARFlex device implanted, seventy-three had a fake operation with no device implanted, and three people in each group stopped having migraines. The NMT STARFlex device made no difference at all. This is not a statement of opinion, and there are no complex statistics involved.

This might also be a good point to mention that the journal *Circulation* had to publish a lengthy correction for the MIST trial because the original paper failed to mention, for example, that Wilmshurst had declined to be listed as an author over concerns about how the study was conducted; that two of the devices were lost in patients' bodies during the procedure (one embolised to the right atrium, one to the left pulmonary artery, both worrying, both were luckily able to be retrieved); and so on.

Back to NMT's three positive case studies with their smiling studio photographs. They were all (it explains in the 2005 NMT annual report) treated with the STARFlex device in the MIST trial. Jean Richards says: 'I feel so much better now. I don't live in fear of a migraine coming on all the time.' Zoe Willows says: 'People at my new job have never known me to have a migraine. I'm a totally different person.'

There are several problems here. Firstly, two of these patients, it seems, are advertising devices they were not treated with. Jean is smiling and advertising CardioSEAL, a successor to the failed STARFlex device, although she was not treated with CardioSEAL; and Liz is advertising BioSTAR, but she was not treated with BioSTAR. I asked NMT why these patients were advertising products with which they were not treated. NMT declined to answer.

Secondly, the patients' anecdotal experiences are entirely misleading: the MIST trial was negative (though I can find no mention of its final results anywhere on the NMT site, which is odd, because it's the only published trial I'm aware of that tests whether NMT's device prevents migraine).

But lastly, the protocol for the MIST trial, as is standard, states that the sponsoring company is not supposed to have access to individual patients. How did NMT get hold of these patients?

I tried to contact Dr Michael Mullen, previously of the Royal Brompton Hospital, now of University College London Hospitals, cardiologist on the MIST trial, to see if he knew how these patients hit the public domain, since the RBH website has a page – hurriedly removed since I contacted them – stating that the MIST trial results were positive (these appear to be initial results, from before the final paper was published), and also featuring the patient Zoe Willows saying 'I've now been completely cured.' Dr Mullen himself appears in a smiling studio shot on the NMT website, and in 2008 declared owning shares in NMT. I invited him to criticise NMT's use of misleading patient anecdotes. He declined. I asked if he knew how the company got hold of the patients, or how these positive results appeared on the RBH website. He said he could not remember.

So I asked NMT. It told me that all three patients got in touch with the medical-device company themselves, spontaneously. I asked NMT if the three patients whose migraines stopped after the fake operation had also got in touch to express their gratitude, because they might be able to provide useful and less misleading anecdotes. NMT declined to answer.

I could then have asked Dr Andrew Dowson, the new lead investigator on MIST, whose licence to practise was restricted by the GMC at the time of the MIST trial, as he had been found guilty of research misconduct in an earlier clinical trial. But by then I was exhausted, and not sure it was worth it.

Meanwhile, NMT's share price has fallen from $20 to 20 cents over four years, perhaps unsurprisingly after the negative results of the MIST trial. A judge has now insisted that the company puts £200,000 into a UK account in case it loses its libel case against Dr Wilmshurst, or the case will be struck out next month, but NMT's solicitor argues that the company's financial situation is 'dire'. This suggests that even if Dr Wilmshurst successfully defends himself, he may never get his £100,000 back. I'm not convinced that a libel law which allows a company like NMT to do this to one man is in society's best interests.

'We Are More Possible Than You Can Powerfully Imagine'

Guardian, 29 July 2009

Today the Australian magazine *Cosmos*, along with a vast number of other blogs and publications, reprinted an article by Simon Singh, in slightly tweaked form, as an act of solidarity. The British Chiropractic Association has been suing Singh personally for the past fifteen months, over a piece in the *Guardian* in which he criticised the BCA for claiming that its members could treat children for colic, ear infections, asthma, prolonged crying, and sleeping and feeding conditions, by manipulating their spines.

The BCA maintains that the efficacy of these treatments is well documented. Singh said the claims were made without sufficient evidence, described the treatments as 'bogus', and criticised the BCA for 'happily promoting' them. At a preliminary hearing in May to decide the meaning of this article, Mr Justice Eady ruled that Singh's wording implied that the BCA was being deliberately dishonest. Singh has repeatedly been clear that he never intended this meaning, but has been forced to defend this single utterance, out of his own pocket, at a cost that has run to six figures.

Soon we will get to the story of the backlash, but first, while you may view this as a free-speech issue, there are also some specific worries raised when people sue in medicine and science.

It is possible in healthcare to do great harm, while intending to do good, and so medicine thrives on criticism: this is how ideas improve, and therefore how lives are saved. The three most highly rated articles in the latest chart from the *British Medical Journal* are all highly critical of medical practice. Academic conferences are often bloodbaths. To stand in the way of ideas and practices being improved through critical appraisal is not just dangerous, it is disrespectful to patients, and even if someone has been technically defamatory in their wording, it is plainly undesirable for all critical discourse in healthcare to be conducted in a stifling climate of fear. Neither the General Medical Council nor the British Medical Association has ever sued anyone for saying that their members are up to no good. I asked them. The idea is laughable.

But beyond whether it is right, there is the more entertaining issue of whether it was wise, and here it is hard to contain a sense of *Schadenfreude* as the chiropractors' world unravels. First, there is the media exposure. This case, and the chilling effects of libel threats in science, have now been covered by *The Times*, the *Daily Mail*, the *Daily Telegraph*, the *Independent*, *Nature*, the *Economist*, *Times Higher Education*, the *Sunday Times*, Channel 4, the *Financial Times*, the *Wall Street Journal*, *Private Eye*, the *Observer*, the BBC, and an editorial in the *British Medical Journal*, to name just a few. This story has travelled around the world.

Most of these articles and programmes drew attention to the evidence for chiropractic's efficacy, which is often not compelling. Some discussed chiropractic's dubious origins: it was invented by a magnet therapist, convicted of practising medicine without a licence, who suddenly decided in 1895 that 95 per cent of all diseases are caused by displaced vertebrae, and compared himself to Christ, Mohammed and Martin Luther. Who knew?

An international petition against the BCA has been signed by professors, journalists, celebrities and more, with Ricky Gervais and Stephen Fry alongside the previous head of the Medical Research Council and the last government Chief Scientific Adviser. There have been public meetings, with stickers and badges. But it is a ragged band of science bloggers who have done the most detailed work. Fifteen months after the case began, the BCA finally released the academic evidence it was using to support its specific claims. Within twenty-four hours this was meticulously taken apart by bloggers, referencing primary research papers and looking in every corner.

Professor David Colquhoun of UCL pointed out, on infant colic, that the BCA cited weak evidence in favour of its claims, while ignoring strong evidence contradicting them. He posted the evidence and explained it. LayScience flagged up the BCA selectively quoting a Cochrane review. Every stone was turned by Quackometer, APGaylard, Gimpyblog, EvidenceMatters, Dr Petra Boynton, MinistryofTruth, Holfordwatch, legal blogger Jack of Kent, and many more. At every turn they have taken the opportunity to explain the principles of evidence-based medicine – the sin of cherry-picking results, the ways a clinical trial can be unfair by design – to an engaged lay audience, with clarity as well as swagger.

Then the formal complaints began. There have been successes with the Advertising Standards Authority, including one in which the ASA concluded that the BCA's claims to treat colic breached the guidelines on 'truthfulness' and 'substantiation'. This interested many, since treating colic was one of the claims over which the BCA sued Singh when he called it 'bogus'.

Professional complaints followed in May, mostly about individual chiropractors' claims. Then, in June, blogger Simon Perry found the BCA database of 1,029 members online, containing four hundred website URLs. He wrote a quick computer program to automatically identify all the chiropractors in the UK who claimed to treat colic, locate their local Trading Standards office, and report them (more than five hundred in total) automatically, followed up with printed letters.

Chiropractic is a profession regulated by the General Chiropractic Council, supervised by the Health Professional Council, which is obliged to investigate all complaints. So Perry reported over five hundred chiropractors to them, alleging that they had made claims without adequate evidence. The GCC rejected his letter, saying that it only considered individual complaints. A pile of individual complaint letters was instantly generated, printed, stuffed into envelopes, and delivered to its door. Astonishingly, ZenosBlog had done exactly the same thing. These thousand complaints are now being investigated.

You may view this as bullying individuals, and initially I had some sympathy for them. But my heart was hardened as a result of reading commentary from the chiropractic and alternative-therapy community saying that Singh must expect six-figure consequences for criticising them, and transgressing the letter of the law, even in just one article.

Some clue as to whether chiropractors feel able to defend these complaints about the evidence for their practices came a few days later. On 8 June the McTimoney Chiropractic Association sent a confidential email to its members, which has been obtained and is available in full on Quackometer. 'If you have a website,' this email begins, 'take it down NOW . . . REMOVE all the blue MCA patient information leaflets, or any patient information leaflets of your own that state you treat whiplash, colic or other childhood problems in your clinic . . . IF YOU DO NOT FOLLOW THIS ADVICE, YOU MAY BE AT RISK FROM PROSECUTION. Finally, we strongly suggest you do NOT discuss this with others' – and on this it was clear – 'especially patients.'

The MCA says the complaints are a 'vexatious campaign against the profession', that it has nothing to hide, and that it believes its members have not intentionally breached any rules regarding their websites' content. The entire MCA website disappeared on the same day, and continues to be nothing more than a holding page (it is 'currently being updated'), but its former site, along with every single chiropractor's website, has been archived in full online by the science blogging community, for anyone who is interested to look.

We could go on, but there are lessons from this debacle – beyond the ethical concerns over suing in the field of science and medicine – and they are clear. First, if you have more reputation and superficial plausibility than evidence to support your activities, then it may be wise to keep under the radar, rather than start expensive fights. But more interestingly than that, a ragged band of bloggers from all walks of life has, to my mind, done a better job of subjecting an entire industry's claims to meaningful, public, scientific scrutiny than the media, the industry itself, and even its own regulator. It's strange that this task has fallen to them, but I'm glad someone is doing it, and they do it very, very well indeed.

Science Is About Embracing Your Knockers

Guardian, 13 November 2010

If science has any credibility, it derives from transparency: when you make a claim about how something works, you provide references to experiments, which describe openly and in full what was done, in enough detail for the experiment to be replicated.

You explain what was measured, and how. Then people can freely discuss what they think this all means in the real world.

Maria Hatzistefanis is a star of lifestyle pages and the owner of Rodial, a cosmetics company which sells a product called 'Boob Job', which it claims will give you a 'fuller bust', 'increase the bust size' and 'plump up the décolleté area' with 'an instant lifting and firming effect', and an increase of half a cup size in fifty-six days. Or rather, an increase of '8.4 per cent'. It's all very precise.

Now, I'm not going to lose a great deal of sleep over anybody who buys a magic cream to make their breasts grow bigger. What worries me is that Maria Hatzistefanis's company is now threatening a doctor with a libel case, simply for daring to voice doubts over these claims.

This is her crime. Dr Dalia Nield told the *Mail* it was 'highly unlikely' that the cream would make your breasts bigger, and questioned the amount of information provided by Rodial. 'The manufacturers are not giving us any information on tests they have carried out. They are not telling us the exact ingredients in the product and how they increase the size of the breast.'

That's fair. I don't trust claims without evidence, especially not unlikely ones about a magic cream that makes your breasts expand. Maybe it does work – I don't particularly care either way – but when I asked the company to send me any evidence it had, or any information on ingredients, it flatly refused to send me anything at all.

This is odd, since I've seen the letter that Rodial's lawyers sent, and they tell Dr Nield: 'Our client on request would have provided all information required on clinical assessment and product ingredients.'

Apparently not.

Dr Nield went on to speculate that the gel could be 'potentially dangerous . . . it may even harm the skin and the breasts – we need a full analysis'. Again, this is perfectly reasonable: anything that has real effects on the body may also have unintended side effects. That is an entirely uncontroversial statement, especially when important information is being withheld.

But then the story gets stranger. When Sense About Science, which has helped drive the campaign for libel law reform in the UK, put out a press release about Rodial threatening Dr Nield with libel, they themselves were contacted by Hegarty LLP, solicitors acting for Rodial Limited. This time they seemed to be trying to stop people from even daring to talk about the existence of their libel threat.

People often ask if there are short cuts in spotting nonsense. In reality, it's not easy to do in a checklist, because there are so many elaborate ways to distort evidence, but for me there is one very clear risk factor. The entirety of science is built on transparency, on giving your evidence and engaging with legitimate criticism. If you hear of a company refusing to hand over the evidence it says supports its claims – whether it is a drug company, or some dismal cosmetics firm – all you know is that you are being deprived of information, and that vital parts of the picture are missing. If you hear that someone is threatening to sue their critics, again, all you know is that people will be intimidated from raising legitimate concerns, and again, you are being deprived of information.

Meanwhile, Dr Nield is now one individual facing a large company. Individual doctors and scientists are commonly asked for their opinion on whether or not medical interventions work. It's plainly in our collective interest that they give honest answers, without fear that their lives will be taken over for years on end by a major company with money and a distorted sense of reputation. It's obviously unhelpful, for society, that people risk losing a house's worth of money even if they successfully defend a libel case, as has happened so many times recently.

With the law in its current state, doctors and scientists might be wise simply to stop giving any view about any drug or other health-related product that is marketed for commercial purposes, in any forum, and make it clear that from now on, decisions about efficacy should be made solely by the manufacturers. Good luck with that.

The Return of Dr McKeith

Guardian, 19 July 2010

What do you do, as a campaigner for libel reform, when a litigious millionaire calls you a liar? This ethical quandary was presented to me last week when the Twitter account of Gillian McKeith – or to give her full medical title, 'Gillian McKeith' – called my book *Bad Science* 'lies'.

Now, firstly, there is little doubt that this is actionable, and probably undefendable. 'Lies' – from personal experience of trying – is one word you can never use in England. Even if you can show that someone is obviously wrong, even if you can show that they probably knew they were wrong, you still need to show that they deliberately distorted the truth, and that's almost always impossible, without direct access to their thoughts. They might just have been mistaken, after all. Or sloppy. Or stupid.

So, for my pleasure, I have a strong case against the litigious millionaire. And I have a reasonably good reputation for honesty to defend. And although I believe libel laws stifle debate in science at great risk to public health, there's no issue of science here.

But I've always believed that in most cases a simple correction, with the same prominence as the initial libel, should be sufficient. That's why I contacted @gillianmckeith, firstly to explain that I'd be happy to debate my concerns about her work, and secondly, to make a simple request: could she please just tweet 'Bad Science by Ben Goldacre is not lies'. That would be fine with me.

But by now all hell had broken loose. @gillianmckeith's Twitter feed was filled with abuse to a random passing tweeter, and rambling, detailed tweets from McKeith explaining how her PhD from a non-accredited correspondence-course college was entirely valid. Then they all disappeared. Then the tone shifted: instead of first-person stuff about Gillian's life, lots of third-person PR tweets appeared. Then they disappeared. Then, as over a thousand people were tweeting about her, making it the top trending topic on Twitter, @gillianmckeith announced, 'Do you really believe this is real twitter site for the GM?'

Yes, replied the geeks. Well, the Twitter account @gillianmckeith is linked to from gillianmckeith.info, explained some. Then that link was deleted. But, explained others, only half-deleted. If you look at the 'source code' for the page, the link is still there, just temporarily inactivated. And that Twitter account is still linked from gillianmckeith.tv, Gillian's YouTube page, and in fact her whole empire. Yes, we really do believe this is the real Twitter site for the real Gillian McKeith. If you're going to play silly buggers online, at least do it competently. And really, very seriously, don't call investigative journalists liars. You never know: we might sue too.

QUACKS

The Noble and Ancient Tradition of Moron-Baiting

Guardian, 29 May 2010

This week a man called Martin Gardner died, aged ninety-five. His popular maths column in *Scientific American* (and fifty books on the subject) spanned the decades, but in 1952 he published a book, *In the Name of Science*, on pseudoscience, quacks and credulous journalists. How much do you think has changed over sixty years?

Immanuel Velikovsky had just published his best-selling book *Worlds in Collision*, explaining how a comet which flew out of Jupiter, and zipped past the earth twice, had then caused the earth to stop spinning, so that the Red Sea would part at precisely the moment when Moses held out his hand. Cars and planes, he explained, are propelled by fuel refined from 'remnants of the intruding star that poured fire and sticky vapour' on the earth. Several years later the comet returned: a precipitate of carbohydrates that had formed in its tail fell to earth in the form of Manna which kept the Israelites fed for forty years.

The science editor of the *New York Herald Tribune* called this book 'a magnificent piece of scholarly research'. But while the correspondents of *Reader's Digest* and *Harper's Magazine* heaped praise upon Velikovsky, his publishers received a flood of letters from scientists. A boycott was organised of all their academic textbooks, the editor who commissioned the book was sacked, and Velikovsky moved to Doubleday, which had no textbook imprint to worry about (and was delighted to have a best-seller).

This was an era when serious people took bullshit more seriously. While today homeopathy is taught in universities eager to serve popular demand, the most notable predecessor to Gardner's *Fads and Fallacies* was *Higher Foolishness*, written in 1927 by David Starr Jordan, the first president of Stanford University. The American Medical Association campaigned hard against press publicity for quacks, and bullshit seemed more pressing. There were signs of a relapse into religious fundamentalism, driven in part by bizarre beliefs such as Velikovsky's, and the indulgence of pseudoscience was playing its part, live and in colour, in some very bad situations.

The bizarre racial theories of the Nazi anthropologists were fresh in the memory, and in Russia things were little better. During the 1930s, communism had turned its back on evolution and Mendelian inheritance, preferring the theories of Trofim Lysenko on the inheritance of acquired characteristics, which sat better with its notions of heritable self-improvement. Sadly, Lysenkoism ran contrary to the experimental evidence, and could only be maintained by sending Russia's geneticists to die in Siberian labour camps, so that by 1949 Russian children were being taught that the revolution had shattered the hereditary structure of the Soviet people, with each generation growing up finer than the last as a result.

But alongside concrete outcomes like death camps, Gardner never loses sight of the parallel tragedy. *Harper's Magazine* – notable for its recent promotion of Aids denialism – was then pushing Gerald Heard's book *Is Another World Watching?*, which explained that tiny flying saucers have visited earth, piloted by two-inch super-intelligent bee people from the planet Mars. At a time when the shelves were filled with magazines called things like *Life*, *True* and *Doubt*, a widespread passion for knowledge was being regularly derailed into nonsense.

A QUACK IN THE RIGHT PLACE;
Or, What we Should Like to See.

So, Gardner has the same fun we have with the homeopaths (while complaining that Marlene Dietrich is a fan), the vitamin-pill peddlers, the anti-vaccination campaigners and the chiropractors, and above all captures their character, which endures: the self-imposed isolation from the corrective of academic criticism, the persecution complex, the grandiosity, the denouncement of critics as being in the pay of darker forces, and the enjoyment of jargon like 'electroencephaloneuromentimpograph', a machine devised by the son of the founder of chiropractic.

I have a copy of the first edition of *In the Name of Science* (they're cheap), but subsequent editions are much more desirable, because they include a supplementary introduction where Gardner takes delight in his hate mail, and especially the mutual indignation that each target expresses at being unfairly associated with the others, whom they regard as the true charlatans.* In sixty years nothing has changed. The best we can hope for is the simple, enduring pleasure of baiting morons.

---

*From the preface to the second edition: 'The first edition of this book prompted many curious letters from irate readers. The most violent letters came from Reichians, furious because the book considered orgonomy alongside such (to them) outlandish cults as dianetics. Dianeticians, of course, felt the same about orgonomy. I heard from homoeopaths who were insulted to find themselves in company with such frauds as osteopathy and chiropractic, and one chiropractor in Kentucky "pitied" me because I had turned my spine on God's greatest gift to suffering humanity. Several admirers of Dr. Bates favoured me with letters so badly typed that I suspect the writers were in urgent need of strong spectacles. Oddly enough, most of these correspondents objected to one chapter only, thinking all the others excellent.'

How Do You Regulate Wu?

Guardian, 20 February 2010

You might have read about the case of Ying Wu this week: a fully qualified traditional Chinese-medicine doctor operating out of a shop in Chelmsford who for several years prescribed high doses of a dangerous banned substance to treat the acne of senior civil servant Patricia Booth, fifty-eight, reassuring her that the pills were as safe as Coca-Cola. Following this her patient has lost both kidneys, developed urinary tract cancer, had a heart attack, and is now on dialysis three times a week. Judge Jeremy Roberts gave Wu a two-year conditional discharge, saying she did not know the pills were dangerous and could not be blamed, because the practice of traditional Chinese medicine is totally unregulated in Britain, a situation which he suggests should be remedied.

This sounds attractive, and has been loudly welcomed by alternative therapists, who see regulation as the path to legitimacy. It's worth noting, in passing, that we do already have systems in place for dealing with dangerous substances (these pills are banned), false claims on the high street (like the regional Offices of Fair Trading, which chose not to use its powers here), and people prescribing treatments which have both powerful effects and dangerous side effects (like doctors, who make bad calls often enough that it's hard to imagine why you'd want people with weaker training handing out dangerous pills).

But special regulation for alternative therapists raises one very simple problem: it's extremely hard to regulate practitioners who make claims based on faith more than evidence. In such a situation, what is your yardstick for whether a clinical decision was reasonable?

Current attempts at regulation have exposed these contra-
dictions. The Complementary and Natural Healthcare Council
(or OfQuack, as it is affectionately known) has a Code of
Conduct which forbids alternative therapists making claims
without evidence. Blogger Simon Perry complained about
every single reflexologist on its register, on the day they joined,
if they were claiming to treat things like arthritis, infertility,
babies with colic, and so on. All were told off, but the CNHC
decided that their fitness to practise was not impaired, because
its reflexology expert said that the practitioners would have
honestly believed their claims to be reasonable, since they
would have been trained to believe that they could treat these
complaints.

So is the training the problem? The government's review
into regulation of alternative therapists has recommended that
it should be compulsory to have a university degree in alternative therapies, and that universities should run such courses.
And what is taught on these courses? You cannot know, because
the universities have gone to shameful lengths over many years,
to the point of multiple appeals at the highest level with the
Information Commissioner, to keep the contents of these
science degrees a closely guarded secret.

I and Professor David Colquhoun of UCL have obtained
occasional course materials from students themselves, who
thought they were going to be taught the scientific evidence base
for alternative medicine, and have been dismayed by what they
received. You can see why the universities wanted to hide them.
Handouts from the Bachelor of Science degree in Chinese
Medicine at Westminster University, for example, show students
being taught – on a science degree – that the spleen is 'the root
of post-heaven essence', 'houses thought (and is affected by
pensiveness/over thinking)' and is responsible for the 'transfor-
mation of qi energy', 'keeping the muscles warm and firm'.

'Marrow helps fill the brain'. 'Sin Jiao assists the lungs'

"dispersing function", spreading fluids to skin in form of fine mist or vapour (so it helps regulate fluid production . . .)'. We also see the traditional anti-vaccine rhetoric – a core marketing tool for alternative therapists – as students are taught that vaccination is a significant cause of cancer.

One lecture by Niki Lawrence on 'Herbal Approaches for Patients with Cancer', meanwhile, discusses the difficulties of the Cancer Act, which was specifically designed to protect patients from the more dangerous extremes of alternative ther-apists' self-belief. 'Legally you cannot claim to cure cancer,' it begins, on a slide headed 'Cancer Treatment and the Law'. 'This is not a problem because: we treat patients not diseases.' Niki then romps on to explain that poke-root is 'especially valuable in the treatment of breast, throat and uterus cancer', *Thuja occidentalis* is 'indicated for cancers of possible viral origin, e.g. colon/rectal, uterine, breast, lung', and *Centella asiatica* 'inhibits the recurrence of cancer'.

It is a tragedy that someone has contracted a fatal condition and is on dialysis. What worries me is that when you try to slot the square peg of fanciful overclaiming and faith-based medi-cine into the round hole of serious regulation and university teaching, you create more problems and confusion than you started with.

Blame Everyone But Yourselves

Guardian, 26 July 2008

Like the practitioners of many professions that kill with some regularity, doctors have elaborate systems for seeing what went wrong afterwards. The answer is rarely 'Brian did it.' This week the papers have been alive with criticism of quack nutritionism after the case of Dawn Page, a fifty-two-year-old mother of two who ended up being treated on intensive care, with seizures brought on by sodium deficiency, and left with permanent brain damage. She had been following the advice of 'nutritional therapist' Barbara Nash. Ms Nash denies liability. Her insurers paid out £810,000.

I will now defend the nutritional therapist Barbara Nash.

There is no doubt that people who declare themselves to be healthcare practitioners are a risk, by virtue of their sheer, uncalibrated self-belief. It must take strong nerves to tell a customer, as they follow 'the Amazing Hydration Diet' – dramatically increasing water intake, and reducing salt intake – that their uncontrollable vomiting is simply 'part of the detoxification process'. Perhaps it was done with the reassuring tones of a clinician. In fact, Mrs Page's lawyers explained, at this point she was told by Ms Nash to increase her water intake to six pints a day.

But I put it to the kangaroo court of the international news media – since this story has now spread as far as America and Australia – that Barbara Nash's confidence in her own judgement cannot be viewed outside its social context.

After completing the rigorous training at the 'College of Natural Nutrition', anyone would naturally believe themselves to be appropriately qualified, and able to give advice confidently. That is certainly the impression I have from reading the college's website. Barbara Nash's confidence in her own abilities seems entirely congruent with that world view. This college operates legally and is well promoted.

Then there are the professional bodies. They have been rather keen to distance themselves from Barbara Nash. In the *Daily Telegraph*, for example: 'The British Association for Applied Nutrition and Nutritional Therapy (BANT) which has its own code of conduct, said Mrs Nash was not a member.' This is not the entire truth. Barbara Nash is advertised on yell.com as a member of BANT. In fact, she was indeed a member of BANT, until last year.

Membership of BANT carries such privileges as 'a listing in the BANT Directory of Practitioners, which is available to the public, and entry on the BANT website', and 'acknowledgement of professional status by the Nutritional Therapy Council'. Endorsed in this way by these bodies, Barbara Nash has every reason to hold her own clinical abilities in high regard. The episode with Dawn Page on intensive care occurred in 2001. These honours were conferred upon her by BANT in 2005.

And of course we should not forget the wider social context: food has become the *bollocks du jour*, with no regard for accuracy whatsoever. This month, the *Daily Telegraph* was printing advice from a self-declared nutrition therapist on folic acid in pregnancy that may actually increase the risk of disabling neural-tube defects in babies, in the same week that it ran a news story telling women that red wine prevents breast cancer, when actually it increases it; and the sofas of daytime television are filled with self-declared nutritionists, because they give us what we want to hear: technical, complicated, sciencey-sounding health advice.

Looking at Barbara Nash's website, I see she carries testimonials from her own appearances on ITV Central's *Shape Up for Summer* slot: 'When I met Barbara [who was the nutritionist for this programme] I wasn't really sure how her eating plan would help me . . . However, it did involve one aspect that I found very difficult to follow, drinking four pints of water a day. I would be the first person to say that I was sceptical but as I had volunteered to take part, I felt that I at least owed it to everyone to try. Was I surprised by the results!'

Promoted, endorsed, trained and buoyed, Barbara Nash has every good reason to think that what she is doing is sensible and correct. Dawn Page – for all that you might think, in an unkind moment, that she was a little gullible – similarly had every reason to believe that Nash was competent. Their view on Nash's competence, and everyone else's, is quite reasonably reinforced by the College of Natural Nutrition, the British Association of Nutritional Therapists, Central TV, and every single journalist, editor, commissioner and producer who has shepherded the bizarre world of made-up nutritional nonsense into our lives.

The specific harm done in this one episode is tragic. It always is. The real measure of professionalism is how you investigate, and what you change. No system would be perfect, but in this case, everyone is queuing up to hold out Barbara Nash as solely responsible. When you miss the real cause, you can be sure that the problem will rise again.

MAGIC BOXES

ADE 651: WTF?

Guardian, 14 November 2009

It's always interesting when people take pseudoscience out of its natural habitat – Islington – and off into a place where the stakes are quite high. Like the polio vaccine scare in Nigeria. Or Aids denialism in South Africa. Or detecting bombs in Iraq, where the *New York Times* and magician James Randi have uncovered some nonsense of epic proportions.

A British company called ATSC is selling a device that can detect guns, ammunition, bombs, drugs, contraband ivory, and truffles. The bomb-detection equipment that you see in airports weighs several tons, and can only operate over tiny distances. The ADE 651 uses 'electrostatic magnetic ion attraction' and can detect these things from a kilometre away, through walls, under the ground, underwater, or even from an aeroplane five kilometres overhead.

ATSC's device is pocket-sized and portable. You simply take a piece of plastic-coated cardboard, which has been through 'the proprietary process of electrostatic matching of the ionic charge and structure of the substance' you're trying to find, pop it into a holder connected to a wand, and start detecting. It's a bit Fisher Price. There are no batteries and no power source: you hold the device to 'charge' it with the energy of your body, and become perfectly relaxed, with a steady pulse and blood pressure. Then you walk with the wand at right angles to your body. If there is a bomb on your left, the wand will drift to the left, and point at it. Like a dowsing rod.

Similar devices have been tested repeatedly and shown to perform no better than chance. No police force or security service anywhere in the developed world uses them. But in 2008 the Iraqi government's Ministry of the Interior bought eight hundred of these devices – the ADE 651 – for $32 million. That's $40,000 each, and they've ordered a further shipment at $53 million. These devices are now being used at hundreds of checkpoints in Iraq, to look for bombs.

Dale Murray, head of the National Explosive Engineering Sciences Security Center at Sandia Labs, which does testing for the US Department of Defense, has tested various similar devices, and they perform at the level of chance. On Tuesday, two people working for the *New York Times* went through nine Iraqi police checkpoints which were using the device, and none found the rifles and ammunition they were carrying (with licences).

Major General Jehad al-Jabiri is head of the Iraqi Ministry of the Interior's General Directorate for Combating Explosives. 'I don't care about Sandia or the Department of Justice or any of them,' he says. 'Whether it's magic or scientific, what I care about is it detects bombs.'

How would you know? There are no independent tests of the ADE 651 that I can find. The simplest explanation is that nobody could really be bothered. Magician James Randi can. He has carried a cheque for $1 million in his jacket pocket for many years, in an admirably expensive act of passive aggression, and he will give this cheque to anyone who can provide proof of supernatural phenomena. Last year he invited the manufacturers of the ADE 651 to come forward, and see if their device works better than chance. They have not.

If you've trousered $85 million, you probably don't care about the Amazing Randi and his puny cheque. We all have our excuses. General al-Jabiri, meanwhile, challenged an *NYT* reporter to test the ADE 651, placing a grenade and a machine pistol in plain view in his office. Every time a policeman used it, the wand pointed at the explosives. Every time the reporter used the device, it failed to detect anything. 'You need more training,' said the General.

As of June 2014 James McCormick – the man behind the company selling this device – has been convicted of fraud and sentenced to ten years, al-Jabiri is in jail for taking bribes from McCormick, and the device is being used in Iraq, Pakistan, Kenya and Lebanon.

After Madeleine, Why Not Bin Laden?

Guardian, 13 October 2007

Danie Krügel is an ex-policeman in South Africa who believes he can pinpoint the location of missing people anywhere on the map. He does this by using his special magic box, which works through something to do with 'quantum physics', but you aren't allowed to know any more than that: these are 'complex and secret science techniques', driven by a 'secret energy source' driving a 'matter orientation system machine'. By simply popping a strand of the missing person's hair – or some other source of DNA – into his box of tricks, Krügel can pinpoint that person's location, anywhere.

This might sound ridiculous to you – or rather, it might sound like the familiar nonsense from psychics, who often involve themselves in cases of missing children – but this week both the *Telegraph* and the *Observer*, as well as several tabloids, featured Krügel in serious news stories on the hunt for Madeleine McCann.

'Traces of Madeleine McCann's body were found on a Portuguese beach weeks after she was reported missing,' said the *Observer*, under the headline 'Forensic DNA Tests "Reveal Traces of Madeleine's Body on Resort Beach"'. The *Telegraph* emphasised the science: 'it emerged the couple had used a scientist to help look for the missing four-year-old using a DNA-tracking device'.

So what do we make of this box, using a new top-secret quantum theory about 'matter orientation', invented and manufactured by a retired police officer currently working as head of health and safety at a university? (Or as one newspaper had it: 'Krügel, of the University of Bloemfontein'.)

This device would have to analyse the DNA at a very high level of resolution, and not just identify DNA, but also the DNA specific to one person, from a distance. It would then use some kind of technology to locate more of that DNA, using a map.

The military applications alone would be incredible. We could use it to find Osama bin Laden, every house burglar in Britain, and Lord Lucan. It would win a Nobel Prize, and the million-dollar prize offered by magician James Randi for anyone who can demonstrate paranormal powers. Why not use the device to locate Randi, and then claim his million?

I rang Krügel to ask him. Are his powers paranormal? He says no. He made a discovery while experimenting with some off-the-shelf electronic devices. I asked if I could see the device: sadly, the answer is no. I asked him what he measured, how he knew he was measuring anything, but he wouldn't say. I asked about the theory, but that's secret. I asked if he had a background in electronics or quantum theory, and he passed. I asked, 'What's a capacitor?' He was offended. I apologised.

Meanwhile, here is Krügel in a South African documentary on his work finding missing children. 'If you get a signature sample of something . . . let's call it organic or non-organic . . . a very small sample. I have developed a method to use that small sample and to create data that I use to search for its origin. So you transmit and you receive.'

'Is there anything metaphysical involved? Are you psychic?'

Krügel: 'I'm a Christian and I put it clearly . . . this is science, science, science!'

Now put yourself, briefly, in the shoes of someone who has lost a child, watching the television through tearful eyes, hoping against hope that your little baby is still alive, not dead, not murdered, not tortured, hoping they will one day be found.

Krügel is gushing: 'Now that's fantastic. To phone the dad and say, "Look, I've got him," or "I have got her. You can come and get him," or "You can come and get her." ' 'How many of those have you had?' asks the interviewer. 'A lot, a lot, a lot.'

Who's Holding the Smoking Gun on Bioresonance?

Guardian, 12 November 2005

Just as swearing is best when old, posh people do it, so bad science is best when it's on BBC News. The video is online, but I'll transcribe. This is a story, delivered with all the authority of television news, about the 'bioresonance' treatment to help people give up smoking. 'The bioresonance treatment is analysing the energy-wave patterns in Jean's body,' it begins. 'It finds the frequency pattern of the nicotine and reverses it. That in theory neutralises the nicotine's energy pattern, so her body won't crave what's been wiped out.'

Reader John Agapiou, who sent this in, wonders what would happen if this device really did work: 'You'd need to extract the nicotine signal very carefully,' he says. 'You wouldn't want to have any traces of "dopamine" or "haemoglobin" in the recording, and nullify those molecules. Or you'd be in real trouble.'

I'm not sure anyone has ever calculated how many different kinds of molecule there are in the human body, but it must be over a million. So this machine, which looks like a piece of modern hospital equipment, records something through funny little pads attached to the skin, and it can filter out precisely the molecule it's looking for. This is extraordinary signal processing.

The BBC goes on: 'That principle has been used to treat illnesses and allergies. Trying to help smokers quit is a new development. There's still no clinical proof that this works, but the clinic says it treats hundred of smokers every week. And of all those who left their cigarettes here over just the last few days, 70 per cent of them will never go back to smoking.'

This is a better success rate for smoking cessation than any other intervention that has ever been studied, including medication, hypnosis, gum, patches and group interventions.

But of course, BBC policy requires balance: there must be someone in the story to question these outlandish claims. This comes in the form of Simon Martin, from *Complementary and Alternative Therapy* magazine. 'If you get a really good machine, with a well-educated, good, ethical practitioner, the sky is the limit really, but there's an awful lot of people out there I think, not very well trained, using inferior equipment, and the sort of results they're getting really shouldn't be trusted.' This is a news story, repeated several times, on BBC television.

AIDS

House of Numbers

Guardian, 26 September 2009

This week, listening to the Guardian Science podcast, I had a treat. Caspar Melville, editor of *New Humanist* magazine, leader of something called the Rationalist Association, had been to see two films at the Cambridge Film Festival. One was a dreary creationist movie that famously misrepresented the biologists interviewed for it. This was obvious bad science, he explained. But the other was different: *House of Numbers*, a new film about Aids, really had something in it. I have now seen this film. It presents itself as a naïve journey by one young film-maker to discover the science behind HIV. In reality it's a pernicious piece of Aids denialist propaganda.

All the usual ideas are there. Aids isn't caused by the HIV virus: it's caused by antiretroviral drugs themselves. Or poverty. Or drug use. Diagnostic tools don't work, and Aids is just a spurious basket diagnosis invented to sell antiretroviral medication for a wide range of unrelated problems. The drugs don't work. You're better off without them.

It would take a book to address all this, and that blizzard of claims, perhaps, is the point of the film, with all the rhetorical devices that have been honed by Aids denialists and creationists over decades. It repeatedly overstates marginal internal disagreements about the details of HIV research, for example, to the extent that eighteen doctors and scientists who were interviewed

in it have issued a statement saying that the director was 'decep-
tive' in his interactions with them, that it perpetuates pseudoscience and myths, and that they were selectively quoted to
make it seem as if they are in disagreement and disarray, when
in fact they agree on all the important facts.

At one point there is an extended sequence explaining that
you can't take a picture of the HIV virus; or maybe you can, but
if you can, different scientists disagree on how, and on whether
their method is best. This is an infantile world view, where stuff
only exists if you can easily photograph it; where the internet,
compound interest and magnetism don't exist either.

There is a memorable skit on diagnostic tests, where the
film-maker manages to find one woman working in a marquee
in a shopping centre in Africa giving HIV tests, who accidentally
misinforms him about why she is asking for information on
his health-risk behaviours. In the film, this one slip by a junior
member of staff in a shopping centre becomes a dramatic
exposé: the HIV diagnosis is a tautology, it argues, a basket
diagnosis for sick people of any kind who engage in risk behaviours; the blood test is unreliable, a piece of theatre; and the
diagnosis is only made because the tester has asked if you are
gay or inject drugs.

But people working on the front line of HIV testing are often
told to ask about risk behaviours during a test, because testing
is a great opportunity to educate people on prevention. What's
more – if you're interested in the statistics of testing – knowledge about your pre-test likelihood of having a condition also
helps the tester to correctly interpret any diagnostic test:
because, as we have covered in this column, for terrorist
screening, for predicting violence in psychiatric patients, indeed
for anything, the likelihood of a false positive with any test is
higher where the population prevalence of a condition is low.
In any case, HIV tests are so reliable that in 2007 an HIV-negative
woman won $2.5 million in damages after she was treated for

Aids without a proper diagnosis, since there was no excuse for the mistake that her doctor made.

The show goes on. We see Neville Hodgkinson, the *Sunday Times* health correspondent who drove that paper's denialist reporting in the 1990s. There is Peter Duesberg, who you will remember from when academic publishers Elsevier forcibly withdrew an article by him in one of their journals.

Then there is an interview with Christine Maggiore, who talks about her difficult decision to go against medical advice by refusing Aids medication, and how much better she feels as a result. What the film doesn't tell you, oddly, is that Christine Maggiore's daughter Eliza Jane died of Aids and PCP pneumonia three years ago, at the age of three; and, as I reported nine months ago, Christine Maggiore herself died two days after Christmas 2008 of pneumonia, aged fifty-two: this is finally acknowledged in the last two seconds of the film, at the end of the lengthy credits, in tiny letters.

Do you give idiots a wider audience when you respond to them? Are they marginal and irrelevant? I'd like to believe that they are. But the duping of Caspar Melville (who has since recanted), and the attention-seeking smugness of the Cambridge Film Festival, both suggest otherwise. I'll never know the right way to deal with any of these people, and I'll always welcome advice.

Aids Denialism at the *Spectator*

Guardian, 24 October 2009

A lot of strange stuff can fly in under the claim that you are 'simply starting a debate'. You may remember the Aids denialist documentary *House of Numbers* from three weeks ago. Since then, it has had a fabulous run. The organisers of the London Raindance Film Festival explained that they were proud to show it, and a senior programmer appeared on YouTube saying they had gone through the film at fifteen-second intervals, finding no inaccuracies at all.

That's pretty good for a film which suggests that HIV doesn't cause Aids, but antiretroviral drugs do, or poverty, or drug use; that HIV probably doesn't exist; that diagnostic tools don't work; that Aids is just a spurious basket of symptoms invented to sell antiretroviral drugs; and the treatments don't work anyway.

Here is Fraser Nelson, editor of the *Spectator*, promoting a *Spectator* event next Wednesday where this film will be screened: 'Is it legitimate to discuss the strength of the link between HIV and Aids? It's one of these hugely emotive subjects, with a fairly strong and vociferous lobby saying that any open discussion is deplorable and tantamount to Aids denialism. Whenever any debate hits this level, I get deeply suspicious.'

Of course people will have concerns. Despite international outcry, from 2000 to 2005 South Africa implemented policies based on the belief that HIV does not cause Aids. The government refused to roll out adequate antiretroviral therapy to their dying population. It has since been estimated in two separate studies that around 350,000 people lost their lives unnecessarily to Aids in South Africa during this period.

'Teach the controversy' is a technique beloved of cranks, from American creationists to anti-vaccination campaigners (with whom Fraser Nelson has also, oddly, flirted). They know that in our modern media, truth is triangulated, halfway between the two most extreme views: doubt alone gets you close to winning.

But debate is also good. So what kind of debate will the *Spectator* be hosting? It advertises a panel of 'leading medical authorities'. There are four people on this panel.

One is Lord Norman Fowler. He is not a 'leading medical authority'.

Charles Geshekter is a Professor of African History from the University of Chicago. He says there is no Aids epidemic in Africa, just poverty, and that belief in the epidemic is a product of racism and 'Western sexual stereotypes'. In fact he calls it 'The Plague That Isn't', and was on President Thabo Mbeki's notorious Aids Advisory Panel in South Africa in 2000.

Beverly Griffin is an emeritus professor at Imperial College, from the field of virology, but not HIV, who is quoted by the virusmyth website as having said in the 1990s that HIV may not cause Aids. Her views may now have changed. I hope they have. I have emailed her, and hope to hear back.

Lastly, Dr Joe Sonnabend is a retired American doctor who was greatly involved in the treatment of people with Aids, but was also long regarded by many in the Aids denialist community as a fellow traveller. He too has said in the past that the link between HIV and Aids is unproven. More recently he has distanced himself from this view.

I'm sure all these people are erudite and accomplished, but this is not a panel of 'leading medical authorities' on the question of whether HIV causes Aids. It's also fair to say that, with the exception of Norman Fowler, all the *Spectator*'s panellists have disputed the mainstream consensus on Aids at one stage or another. I'm not saying that is unacceptable, or presuming their current position. But they may not reflect the overwhelming consensus – no dirty word – that HIV causes Aids, and that antiretroviral medication is an imperfect but overall beneficial treatment.

And then there is the film. We can't rehash its flaws, but I would ask Fraser Nelson about one scene. Christine Maggiore appears throughout, talking emotively, explaining her choice not to take Aids medication, and saying that this is why she is alive.

But Christine Maggiore is dead, Fraser. The film tells you that, but only in tiny letters at the very end, and it says no more. She died of pneumonia, aged fifty-two, and her daughter died of untreated Aids, aged three. Because of her beliefs about Aids, Christine Maggiore did not take medication which has been proven to reduce the risk of HIV transmission to unborn children during pregnancy. Her daughter, Eliza Jane, was not tested for HIV during her short life. Then she died, aged three, of Aids.

Children don't often drop dead aged three. Adults don't often die aged fifty-two. These facts should be front and centre stage, in large, bold letters, scrolling across the screen as Maggiore speaks out passionately against Aids treatments. I can't see how a film like this can possibly be a helpful starting point for an informed debate. It's not 'controversial', it's pointlessly misleading. 'Starting a debate' is fine. With this film, and with these panellists, the *Spectator* has framed a very odd event indeed.

ELECTROSENSITIVITY

Wi-Fi Wants to Kill Your Children . . . But Alasdair Philips of Powerwatch Sells the Cure!

Guardian, 26 May 2007

Won't somebody, please, think of the children? Three weeks ago I received this email from a science teacher. 'I've just had to ask a BBC *Panorama* film crew not to film in my school or in my class because of the bad science they were trying to carry out,' it began, then went on to describe in perfect detail the *Panorama* programme which aired this week. This show was on the suppressed dangers of radiation from wi-fi networks, and how they are harming children. There was no science in it, just some 'experiments' the programme-makers did for themselves, and some duelling experts. *Panorama* disagreed with the WHO expert, so he was smeared for not being 'independent' enough, and working for a phone company in the past. I don't do personal smears. But *Panorama* started it. Independence is clearly very important to them. How independent were the BBC, and the 'experiments' they did?

They had twenty-eight minutes, I have seven hundred words.

In the show, you can see them walking around Norwich with a special 'radiation monitor'. Radiation is their favourite word, and they use it thirty times, once a minute, although wi-fi is 'radiation' in the same sense that light is. 'Ooh, it's well into the red there,' says reporter Paul Kenyon, holding up the detector. That sounds bad.

Well into the red on what? It's tricky to calibrate measurements, and to decide what to measure, and where the cut-off should be for 'red'. *Panorama*'s readings were 'well into the red' on 'the COM Monitor', a special piece of detecting equipment designed from scratch and built by Alasdair Philips of Powerwatch: the man who leads the campaign against wi-fi. His bespoke device is manufactured exclusively for his own outfit, Powerwatch, and he will sell one to you for just £175. Alasdair decided what 'red' meant on *Panorama*'s device. That's not very independent.

Panorama did not disclose where this detector came from. And they know that Alasdair Philips is no ordinary 'engineer doing the readings', because they told us in the show, but they didn't tell the school that, as our science teacher says: 'They wanted to take some measurements in my classroom, compare them to the radiation from a phone mast and film some kids using wireless laptops. They introduced "the engineer", whom I googled.'

He found it was the same man who runs Powerwatch, the pressure group campaigning against mobile phones, wi-fi and 'electrosmog'. As our science teacher explained, this man isn't necessarily very independent. In Alasdair's Powerwatch shop you can buy shielded netting for your windows at just £70.50 per metre, and special shielding paint at £50.99 per litre. To paint just one small, eleven-foot-square bedroom refuge in your house with Powerwatch's products would require about ten litres of this special product, costing you £500.

When the children saw Alasdair's Powerwatch website, and the excellent picture of the insulating mesh beekeeper hat that he sells (£27) to 'protect your head from excess microwave exposure', they were astonished and outraged. *Panorama* were calmly expelled from the school.

So what about *Panorama*'s classroom experiment? It wasn't very well designed, as was pointed out by a classroom full of children at the time. 'They set about downloading the biggest file they could get hold of – so the wi-fi signal was working as powerfully as possible – and took the peak reading during that,' says our science teacher. It was a great teaching exercise, and the children made valuable criticisms of *Panorama*'s methodology, including: 'We're not allowed to download files, so it wouldn't be that strong,' 'Only a couple of classes have wi-fi,' and 'We only use the laptops a couple of times a week.'

Panorama planned to have the man from Powerwatch talk to the students for about ten minutes about how wi-fi worked, and what effects it had on the human body. Then they were going to reveal the readings he had got from the mast, compare them to what Powerwatch had measured in the classroom, and film the kids' reaction to the news. None of this sounds very independent.

'Surprisingly enough, the readings in my room were going to be higher (about three times higher, I believe), and with the kids having been briefed by the engineer from Powerwatch first, they were hoping for a reaction that would make good telly.' Sadly for them, it didn't happen. 'We told *Panorama* this morning that as they hadn't been honest with us about what was going on and because of the bad science they were trying to pass off, we didn't want them to film in the school or with our students.' The images of children you see in the programme are just library footage.

I'm sure there should be more research into wi-fi. If *Panorama* had made a twenty-eight-minute programme about the scientific evidence, we would be discussing that. Instead they produced 'radiation' scares, and smears about whether people are 'independent'. People in glass houses throw stones at their own risk.

A BBC spokesperson said: 'Alasdair Philips is one of a handful of people with the right equipment to do this test. He was only used in this capacity and was not given an opportunity to interpret the readings let alone campaign on them in the film. We filmed the tests taken at the school and didn't return.'

Why Don't Journalists Mention the Data?

British Medical Journal, 16 June 2007

For two years now the British news media have been promoting the existence of a new medical condition, called electrosensitivity, or electromagnetic hypersensitivity. The story – or hypothesis – is that a wide range of symptoms are caused by acute exposure to electromagnetic signals, and that these symptoms are improved when the signal is removed.

The features of this condition include a range of problems which often end up being characterised by doctors as 'medically unexplained symptoms': tiredness, difficulty concentrating, headaches, nausea, bowel complaints, aches in the limbs, crawling sensations or pain in the skin, and more, for which no clear medical explanation is found. Such problems have existed since long before the appearance of 'electrosensitivity', and the absence of a clear cause is frustrating for both patients and doctors.

If these symptoms were caused by electromagnetic signals, then it should prove possible to study that, ideally in double-blind conditions. The media coverage invariably focuses on the scandal of how research into this area has been neglected: but in fact, dozens of double-blind studies have been performed. A typical experiment involves a mobile phone or wireless network device, hidden in a bag or box. Each subject – chosen from people who report that their symptoms are caused by electromagnetic signals – records their symptoms over time, without knowing if the phone is on or off.

There have now been thirty-seven* such double-blind 'provocation studies' published in the academic literature, and they

---

* The thirty-seventh study was released in the fortnight between the previous article and this one.

are almost all negative. Seven studies did find some statistically significant effect for electromagnetic signals: but for two of those, even the original authors have been unable to replicate the results; for the next three, the results seem to be statistical artefacts (they either use one-tailed t-tests, which assume that the effect can only be negative, and so lower the bar of proof either way; or they make multiple comparisons without accounting for that in the analysis); and for the final two, the positive results are mutually inconsistent (one shows worsened mood with provocation, and the other shows improved mood, which makes the one-tailed t-tests in other studies seem even less reasonable).

These studies test the very claim being repeatedly made in the media: that symptoms are brought on by exposure to a source of electromagnetic signals, and cease when the source is removed. Mostly they are ignored. A recent *Panorama* documentary on BBC 1, covering the possible dangers of wi-fi computer networks, went further. A large chunk of the programme was devoted to electrosensitivity, and the programme-makers followed someone into a lab at Essex University, where they had participated in a provocation study. We were told that this subject correctly identified when the signal was present or absent two thirds of the time, against a visual backdrop of laboratory equipment.

But this was anecdote, dressed up as data. The study is currently unpublished. We don't know the protocol, or whether 2/3 for one subject would be statistically significant (there may have been only three exposures in total, for example). We don't know the results of other subjects. But most crucially, there is no mention that this single selected subject – in a single unpublished study – produced a result that seems to conflict with a literature of thirty-seven studies which are completed, published and, overall, negative. Even if this whole Essex study was positive – which seems unlikely – that would still need to be put in context with the dozens of negative findings that exist already.

Why doesn't the media ever mention this data? Perhaps they deliberately leave it out. Perhaps they never came across it, and are incompetent. Or perhaps they simply lifted their stories verbatim, from aggressive and well-coordinated lobbyists, who promote this new diagnosis and, in many cases, sell expensive equipment to sufferers.

There may also be a darker side. Electrosensitivity lobbyists are not simply silent on the provocation studies: many of them launch vicious attacks on anyone who dares to mention this data, saying they are insensitive, that they are attacking sufferers, and – crucially – that they are denying the reality of patients' symptoms. Symptoms, of course, stand as real, regardless of their cause; and if anyone is inflicting harm, it may be those who obfuscate on the causes. It takes bravery, but we can only develop better treatments through better understanding.

POST-MODERNISM

Archie Cochrane: 'Fascist'

Guardian, 19 August 2006

Sometimes you know an academic paper has overplayed its hand just from the title. 'Deconstructing the Evidence-Based Discourse in Health Sciences: Truth, Power and Fascism'– from the current *International Journal of Evidence-Based Healthcare* – is one such paper. Even Rik Mayall in *The Young Ones* might pull back from using the word 'fascist' – or derivatives of it – twenty-eight times in six pages.

Initially I thought it might be a spoof. After all, who could forget the Sokal hoax, where a Professor of Physics at NYU submitted 'Transgressing the Boundaries: Towards a Transformative Hermeneutics of Quantum Physics' to *Social Text*, the leading postmodernist academic journal. This deliberately meaningless joke article – purporting to undermine the entire discipline of physics – was accepted and published, to universal delight.

But this new article is very real. Here's what the authors put in the 'objectives' section of their abstract: 'The philosophical work of Deleuze and Guattari proves to be useful in showing how health sciences are colonised (territorialised) by an all-encompassing scientific research paradigm – that of post-positivism – but also and foremost in showing the process by which a dominant ideology comes to exclude alternative forms of knowledge, therefore acting as a fascist structure.'

297

If I can put my fascist cards on the table, these are not 'objectives'. Setting details aside, here is a quote from their authority figure, French philosopher Félix Guattari, to illustrate the clarity of his thinking: 'We can clearly see that there is no bi-univocal correspondence between linear signifying links or archi-writing, depending on the author, and this multireferential, multidimensional machinic catalysis.' And from Gilles Deleuze: 'In the first place, singularities-events correspond to heterogeneous series which are organized into a system which is neither stable nor unstable [Jesus], but rather "metastable", endowed with a potential energy wherein the differences between series are distributed.'

These characters are being recruited to attack the notion of evidence-based medicine, and the argument of this paper – it's not an easy read – seems to be that: evidence-based medicine rejects anything that isn't a randomised control trial (which is untrue); the Cochrane Library, for some reason, is the chief architect of this project; and lastly, that this constitutes fascism, in some meaning of the word the authors enjoy, twenty-eight times.

Here's a flavour: 'The classification of scientific evidence as proposed by the Cochrane Group [sic] obeys a fascist logic. This "regime of truth" ostracises those with "deviant" forms of knowledge. When the pluralism of free speech is extinguished, speech as such is no longer meaningful; what follows is terror, a totalitarian violence.' They make repeated allusions to Newspeak. At one point they seem to identify epidemiologists with George W. Bush.

Now, firstly, they are plain wrong about the Cochrane Library, an organisation which simply produces systematic reviews of the published medical literature: Cochrane doesn't only use trial data, in fact many Cochrane reviews contain no trials at all. This is pure ignorance.

But there is a more important general issue here. Evidence-

based medicine is often portrayed – especially by ageing professors from the dying era of eminence-based medicine – as
soulless and algorithmic. But that is a foolish caricature. EBM,
in all the key textbooks, from the earliest editorials, is about
using quantitative information alongside all other forms of
knowledge: taking account of clinical judgement, and patients'
wishes, and boring things like the availability of local services.
It does not denigrate other forms of knowledge, like clinical
experience or patient preference: it seeks to augment and inform
them. EBM is not about being an automaton.

That's all a bit sensible. How about some more childish
attacks, ideally involving fascism? OK, then. I will wear their
label of 'fascist' with a cheeky grin. But Archie Cochrane, on
the other hand – pioneering epidemiologist, and inspiration

Archie Cochrane (left) as a captain in the
International Brigade, c.1936.

for the Cochrane Library – might see things a little differently. After the war, and after working on miners' lung disease, he helped to inspire a democratising culture shift towards evidence-based practice throughout the whole of medicine, and as a consequence, he has probably saved more lives than any single doctor you know. Before that, he was a prisoner of war for four years in Nazi Germany ('The main reason for my capture was my inability to swim to Egypt'). And before that, in 1936, he dropped out of medical school and travelled to Spain to join the International Brigade, where he fought genuinely violent totalitarian oppression, the fascists of General Franco, with his own two hands.

Now. What did you do in your summer holidays?

IRRATIONALITY

The Golden Arse-Beam Method

Guardian, 9 July 2011

Since I was a teenager, whenever I have a pivotal life event coming – an exam, or an interview – I perform a ritual. I sit cross-legged on the floor, and I imagine an enormous golden beam of energy coming out of my arse.

I picture this anal beam passing through each layer beneath me: through the kitchen of the flat below, through the shop and its basement, past gas pipes and sewers and then deep into the earth, where it spreads out into a glorious branching root network, sucking power from the earth. I picture this energy surging through me; I visualise the outcome I want, in enormous detail; and I will it to happen, for about five minutes.

Surprisingly enough, this nonsense is broadly supported by data from randomised controlled trials.

One example was published last month. About two hundred students were randomly assigned to four groups, each with activities supposed to increase their fruit intake. The control group just repeated their goal to themselves ('Eat more fruit'). Another group concentrated on elaborate mental images of themselves enjoying fruit. The third repeated verbal plans for specific situations ('When I see fruit, I will . . .'). The last group visualised elaborate scenes of encountering fruit, picking it up, touching it, eating it.

Among participants eating lots of fruit already, four portions a day, there wasn't much change in their subsequent habits. Among people eating less fruit to begin with, one and a half portions a day, everyone increased their intake; but the ones performing the most elaborate mental imagery did so much more (their intake doubled).

It's not a perfect study – I don't like subgroup analyses for a start, and it only followed up participants for seven days – but it's not alone. An earlier study from 2009 randomly assigned a hundred students either to a control group or to a couple of forms of imagery, picturing themselves choosing a healthy snack over an unhealthy one. The imagery group went on to have more healthy snacks.

Meanwhile, a meta-analysis from 2006 collectively analyses the results of ninety-four studies and finds that 'implementation intentions' ('If I am in situation X, I will do Y') had a positive effect overall on goal achievement.

So there's probably something there, and this research tells us some interesting things about science. Firstly, I think this kind of research is useful. Rupert Sheldrake is the researcher who claims dogs can sense their owner is coming home before they arrive. I disagree with him on a lot, but he has one great idea: that each year, a proportion of the research budget – a hundredth, a thousandth – should be spent on whatever the public vote for. Most of it would go on MMR and homeopathy, of course, but some of it might go on testing, revising and improving stuff that improves people's everyday lives.

Secondly, it shows us that even if you're wrong about how something works, it might still work. I was sold the golden-bum-beam stuff with a lot of nonsense about quantum hippy energy, but I've always thought of it as a perfectly sensible way to combat distractibility. Effective things can come from silly places.

Illusions of Control

Guardian, 4 December 2010

Why do clever people believe stupid things? It's difficult to make sense of the world from our own small atoms of experience, and a new paper in the *British Journal of Psychology* this month shows how we can create illusions of causality, much like visual illusions, if we manipulate the clues and cues.

These researchers took 108 students and split them into two groups. Both were told about a fictional disease called 'Lindsay Syndrome', that could potentially be treated with something called 'Batarim'. Then they were told about a hundred patients, slowly, one by one, hearing each time whether the patient got Batarim or not, and whether they got better.

When you're hearing about patients one at a time, in a dreary monotone, it's hard to piece together an overall picture of whether a treatment works (this is one reason why, in evidence-based medicine, 'expert opinion' is ranked as the least helpful form of information). So, while I can tell you that overall 80 per cent of these imaginary patients got better, regardless of whether they got Batarim or not (the drug didn't work) that isn't how it appeared to the participants. They overestimated its benefits, as you might expect; but the extent to which they overestimated its effectiveness depended on how the information was presented.

The first group were told about eighty patients who got the drug, and twenty who didn't. The second group were told about twenty patients who got the drug, and eighty who didn't. That was the only difference between the two groups, but the students in the first group estimated the drug as more effective, while the estimates of the students who were told about only twenty patients receiving it were closer to the truth.

Why is this? One possibility is that the students in the second group saw more patients getting better without the treatment, and so got a better intuitive feel for the natural history of the condition, while those in the other group who were told about eighty patients getting Batarim were barraged with data about people who took the drug and got better.

This is just the latest in a whole raft of research showing how we can be manipulated into believing that we have control over chance outcomes, simply by presenting information differently, or giving cues which imply that skill had a role to play. One series of studies has even shown that if you manipulate someone to make them feel powerful (by remembering a situation in which they were powerful, for example), they imagine themselves to have greater control over outcomes that are determined purely by chance, which perhaps goes some way to explaining the hubris of the great and the good.

We know about optical illusions, and we're familiar with the ways that our eyes can be misled. It would be nice if we could also be wary of cognitive illusions that affect our reasoning apparatus. But more than that, like the 'Close door' buttons in a lift – which, it turns out, are often connected to nothing at all – these illusions are beautiful modern curios.

Empathy's Failures

Guardian, 2 October 2010

Like all students of wrongness, I'm fascinated by research into irrational beliefs and behaviours. But I'm also suspicious of how far you can stretch the findings from a laboratory into the real world. A cracking new paper from *Social Psychology and*

Personality Science makes a neat attempt to address this short-coming.

Loran Nordgren and Mary McDonnell wanted to see whether our perception of the severity of a crime was affected by the number of people affected. Sixty students were given a vignette to read about a case of fraud, where either three people or thirty people were defrauded by a financial adviser, but all the other information in the story was kept the same.

In an ideal world, you'd imagine that someone who harmed more people would deserve a harsher treatment. Participants were asked to evaluate the severity of the crime, and recommend a punishment: even though fewer people were affected, participants who read the story with only three victims rated the crime as more serious than those who read the exact same story, but with thirty victims.

And more than that, they acted on this view: with a maximum possible sentence of ten years, people who heard the three-victim story recommended an average prison term one year longer than the thirty-victim people. Another study, where a food-processing company knowingly poisoned its customers to avoid bankruptcy, gave similar results.

Now, it's nice that two studies were carried out into the same idea, but I always worry about experiments like this, because they demonstrate an effect in the rarefied environment of the lab, while the real world can be much more complicated.

So what's great about this paper is that it has two halves: the researchers went on to examine the actual sentences given in a representative sample of 136 real-world court cases to people who were found guilty of exactly these kinds of crimes, but with different numbers of victims, to see what impact the victim-count had.

The results were extremely depressing. These were cases where people from corporations had been found guilty of negli-gently exposing members of the public to toxic substances such as asbestos, lead paint or toxic mould, and their victims had all suffered significantly. They were all from 2000 to 2009, they were all jury trials, and the researchers' hypothesis was correct: people who harm larger numbers of people get significantly lower punitive damages than people who harm smaller numbers of people. Juries punish people less harshly when they harm more people.

Now, it seems to me that alternative explanations may possibly play a contributory role here: cases where lots of people were harmed may involve larger companies, with more expensive and competent lawyers, for example. But in the light of the earlier experiment, it's hard to discount a contribution from empathy, and this is a phenomenon we all recognise.

When he appeared on *Desert Island Discs*, Rolf Harris chose to take his own song 'Two Little Boys' with him. When the First World War broke out, Rolf explained, his father and uncle had both joined up, his father lying about his younger brother's age so they could both join the fight. But their mother found out and dobbed them in, because she couldn't bear the thought of losing both her sons so young. Rolf's uncle joined up two years later when he came of age, was injured, and died on the front. Rolf's dad was beside himself, and for the rest of his life he believed that no matter what the risks, if he had been in the same infantry group he could have crawled out and saved his younger brother, just like in the song. Rolf played 'Two Little Boys' to his grandmother just once. She sat through it quietly, took it off at the end, and said quietly, 'Please don't ever play that to me again.'*

---

* In 2014 Rolf Harris was found guilty of several sexual offences against children and jailed for over five years. I've left this piece in, partly as a reminder that abusers don't always have the word 'monster' tattooed on their forehead. That phrasing comes from PinkZapCat, when I asked on Twitter about deleting the column with Rolf in it. I think it's extremely wise.

This story always makes me cry a little bit. Two million people die of Aids every year. It never has the same effect.

Blind Prejudice

Guardian, 4 September 2010

Everyone likes to imagine they are rational, fair and free from prejudice. But how easily are we misled by appearances? Noola Griffiths is an academic who studies the psychology of music. This month she's published a cracking paper on what women wear, and how that affects your judgement of their performance. The results are predictable, but the context is interesting. Four female musicians were filmed playing in three different outfits: a concert dress, jeans, and a nightclubbing dress. They were also all filmed as points of light, wearing a black tracksuit in the dark, so that the only thing to be seen – once the images had been treated – was the movement of some bright white tape attached to their major joints.

All these violinists were music students, from the top 10 per cent of their year, and to say they were vetted to ensure comparability would be an understatement: they were all white European, size 10 dress, size 4 or 5 shoe, and aged between twenty and twenty-two. They were even equivalently attractive, according to their score on the MBA California Facial Mask, which is some kind of effort to derive a number denoting hotness, using the best fit of a geometric mask over someone's face. That may well be ridiculous: I'm just saying they tried.

In fact they did better. All the performances were also standardised at 104 beats per minute, so the audio tracks from each musician could be replaced with a recording of a single performance, by someone who was never filmed, for each of the various pieces in the study. This meant there was no room for anyone to argue that the clothes made the musicians perform differently, because the audio was the same for everyone; and when the researchers checked, in a pilot study, nobody spotted the dummy audio track.

Then they got thirty different musicians – a mixture of music students and members of the Sheffield Philharmonic – and sat each of them down to watch video clips with various different permutations of clothing, player and piece. All were invited to give each performance a score out of six for technical proficiency and musicality.

The results were inevitable. For technical proficiency, performers in a concert dress were rated higher than if they were in jeans or a clubbing dress, even though the actual audio performance was exactly the same every time (and played by a different musician, who was never filmed). The results for musicality were similar: musicians in a clubbing dress were rated worst.

Experiments offer small, constricted worlds, which we hope act as models for wider phenomena. How far can you apply this work to wider society? There's little doubt that women are still discriminated against in the workplace, but each individual situation has so many variables that it can be difficult to assess clearly.

The world of music, however, makes a good test tube for bigotry. That's because in the 1970s and 1980s most orchestras changed their audition policy, in an attempt to overcome biases in hiring, and began to use screens to conceal the candidates' identity.

Female musicians in the top five US symphony orchestras gradually rose from 5 per cent in the 1970s to around 25 per cent. Of course, this could simply have been due to wider societal shifts, so Goldin and Rouse conducted a very elegant study (titled 'Orchestrating Impartiality'): they compared the number of women being hired at auditions with and without screens, and found that women were several times more likely to be hired when nobody could see that they were women.

What's more, using data on the changing gender make-up of orchestras over time, they were able to estimate that from the 1970s to 2000 – the era in which casual racism and sexism in popular culture shifted to more covert forms – between 30 per cent and 55 per cent of the trend towards greater equality was driven simply by selectors being forced not to see who they were selecting. I don't know how you'd apply the same tools to every workplace. But I'd like to see someone try.

Yeah, Well, You Can Prove Anything with Science

Guardian, 3 July 2010

What do people do when confronted with scientific evidence that challenges their pre-existing view? Often they will try to ignore it, intimidate it, buy it off, sue it for libel, or reason it away.

The classic paper on the last of those strategies is from Lord in 1979: the researchers took two groups of people, one in favour of the death penalty, the other against it, and then presented each with a piece of scientific evidence that supported their pre-existing view, and a piece that challenged it. Murder rates went up, or down, for example, after the abolition of capital punishment in a state, or were higher or lower than in neighbouring states with or without capital punishment. The results were as you might imagine: each group found extensive methodological holes in the evidence they disagreed with, but ignored the very same holes in the evidence that reinforced their views.

Some people go even further than this, when presented with unwelcome data, and decide that science itself is broken. Politicians will cheerfully explain that the scientific method simply cannot be used to determine the outcomes of a drugs policy. Alternative therapists will explain that their pill is special, among all pills, and you simply cannot find out if it works by using a trial.

How deep do these views go, and how far do they generalise? In a study now published in the *Journal of Applied Social Psychology*, Professor Geoffrey Munro took around a hundred students and told them they were participating in research about 'judging the quality of scientific information'. First, their views on whether homosexuality might be associated with mental illness were assessed, and then they were divided into two groups.

The first group were given five research studies that confirmed their pre-existing view. Students who thought homosexuality was associated with mental illness, for example, were given papers explaining that there were proportionally more gay people than members of the general population in psychological treatment centres. The second group were given research that contradicted their pre-existing view. (After the study was finished, we should be clear, they were told that all these research papers were fake, and given the opportunity to read real research on the topic if they wanted to.)

Then they were asked about the research they had read, and asked to rate their agreement with the following statement: 'The question addressed in the studies summarized . . . is one that cannot be answered using scientific methods.'

As you would expect, the people whose pre-existing views had been challenged were more likely to say that science simply cannot be used to measure whether homosexuality is associated with mental illness.

But then, moving on, the researchers asked a further set of questions, about whether science could be usefully deployed to understand all kinds of stuff, all entirely unrelated to stereotypes about homosexuality: 'the existence of clairvoyance', 'the effectiveness of spanking as a disciplinary technique for children', 'the effect of viewing television violence on violent behavior', 'the accuracy of astrology in predicting personality traits', and 'the mental and physical health effects of herbal medications'.

The students' views on each issue were added together to produce one bumper score on the extent to which they thought science could be informative on all of these questions, and the results were truly frightening. People whose pre-existing stereotypes about homosexuality had been challenged by the scientific evidence presented to them were more inclined to believe that science had nothing to offer – on any question, not just on homosexuality – when compared with people whose views on homosexuality had been reinforced.

When presented with unwelcome scientific evidence, it seems – in a desperate bid to retain some consistency in their world view – many people would rather conclude that science in general is broken.

Superstition

Guardian, 12 June 2010

We all strive to be right. But what if people who are wrong have better lives? This week, a German study in *Psychological Science* appears to show that being superstitious improves performance, on a whole string of different tasks.

To be clear, I'm always conflicted over this kind of psychology research. On my left shoulder there is an angel. She says it's risky to extrapolate from rarefied laboratory conditions to the real world. She says that publication bias in this field is extensive, so whenever researchers get negative findings, they're probably left unpublished in a desk drawer. And she says it's uncommon to see a genuinely systematic review of the literature on these topics, so you rarely get to see all the conflicting research in one place. My angel has read the books of Malcolm Gladwell, and she finds them to be silly and overstated.

On my right shoulder, there is a devil: she thinks this stuff is *cool and fun*. She is typing right now.

The researchers did four miniature experiments. In the first, they took twenty-eight students, over 80 per cent of whom said they believed in good luck, and randomly assigned them to either a superstition activator or a control condition. Then they put them on a putting green. To activate a superstition, for half of them, when handing over the ball the experimenter said: 'Here is your ball. So far it has turned out to be a lucky ball.' For the other half, the experimenter just said, 'This is the ball everyone has used so far.' Each participant then had ten goes at trying to get the ball into the hole from a distance of 100 cm: and lo, the students playing with a 'lucky ball' did significantly better than the others, with a mean score of 6.42, against 4.75 for the others.

Then they moved on to a second experiment. Fifty-one students were asked to perform a motor-dexterity task: an irritating, fiddly game where they had to get thirty-six little balls into thirty-six little holes by tilting a Perspex box. Beforehand, they were randomly assigned to one of three groups, each of which heard a different phrase just before starting. The superstition activator was 'I press the thumbs for you,' a German equivalent of the English expression 'I've got my fingers crossed.' The two 'control' or comparison groups were interesting. One group were told 'I press the watch for you,' with the idea that this implied a similar level of encouragement (I'm not so sure about that), and the other were told, 'On "go", you go.' As predicted, the participants who were told someone was keeping their fingers crossed for them finished the task significantly faster.

Then things got more interesting, as the researchers tried to unpick why this was happening. They took forty-one students who had a lucky charm, and asked them to bring it to the session. It was either kept in the room, or taken out to be 'photographed'. Then they were told about the memory task they were due to perform, and asked a whole bunch of questions about how confident they felt. The ones with their lucky charm in the room performed better on the memory game than those without; but more than that, they reported higher levels of 'self-efficacy', which was correlated with performance.

Finally, they probed these mechanisms even further. Thirty-one students were asked to bring their lucky charm. It was either taken away or not, and they were given an anagram task. Before starting, they were asked to set a goal: what percentage of all the hidden words did they think they could find? Then they began. As expected, participants who had their lucky charm in the room performed better, and reported a higher degree of 'self-efficacy' as before. But more than that, people who had their lucky charm in the room set higher goals, and also persisted longer in working on the task.

So there you go. Almost everyone has some kind of superstition (mine is that I should mention I noticed this study through my friends Vaughan Bell and Ed Yong on Twitter). What's interesting is that superstition works, because it improves confidence, lets you set higher goals, and encourages you to work harder. In a lab. In one experiment. In a field riven with publication bias. You now know everything you need to decide if this applies to your life.

Evidence-Based Smear Campaigns

Guardian, 1 May 2010

Elections are a time for smearing. But do smears work, and if so, what's the best way to combat them? A new experiment published this month in the journal *Political Behavior* tries to examine the impact of corrections. The findings are disturbing: far from changing people's minds, if you are deeply entrenched in your views, a correction will only reinforce them.

The first experiment used articles claiming that Iraq had weapons of mass destruction immediately before the US invasion. One hundred and thirty participants were asked to read a mock news article, attributed to Associated Press, reporting on a Bush campaign stop in Pennsylvania during October 2004. The article describes Bush's appearance as 'a rousing, no-retreat defense of the Iraq war', and gives genuine Bush quotes about WMD: 'There was a risk, a real risk, that Saddam Hussein would pass weapons or materials or information to terrorist networks, and in the world after September the 11th . . . that was a risk we could not afford to take.' And so on.

The 130 participants were then randomly assigned to one of two conditions. For half of them, the article stopped there. For the other half, the article continues, and includes a correction: it discusses the release of the Duelfer Report, which documented the lack of Iraqi WMD stockpiles – and the lack of an active production programme – immediately prior to the US invasion.

After reading the article, subjects were asked to state whether they agreed with the following statement: 'Immediately before the US invasion, Iraq had an active weapons-of-mass-destruction program, the ability to produce these weapons, and large stockpiles of WMD, but Saddam Hussein was able to hide or destroy these weapons right before US forces arrived.' Their responses were measured on a five-point scale ranging from 'strongly disagree' to 'strongly agree'.

As you would expect, those who self-identified as conservatives were more likely to agree with the statement. Separately, meanwhile, more knowledgeable participants (independently of political persuasion) were less likely to agree. But then the researchers looked at the effect of whether you were also given the correct information at the end of the article, and this is where things get interesting. They had expected that the correction would be less effective for more conservative participants, and this was true, up to a point. For very liberal participants the correction worked as expected, making them more likely to disagree with the statement that Iraq had WMD, when compared with those who were also very liberal but who received no correction. For those who described themselves as left of centre or centrist, the correction had no effect either way.

But for people who placed themselves ideologically to the right of centre, the correction wasn't just ineffective, it actively backfired: conservatives who received a correction telling them that Iraq did not have WMD were *more* likely to believe that Iraq had WMD, when compared with those who were given no correction at all. You might have expected people simply to dismiss a correction that was incongruous with their pre-existing view, or to regard it as having no credibility: in fact, it seems such information actively reinforced their false beliefs.

Maybe the cognitive effort of mounting a defence against the incongruous new facts entrenches you even further. Maybe you feel marginalised and motivated to dig in your heels. Who knows? But these experiments were then repeated, in various permutations, on the issue of tax cuts (or rather, the idea that tax cuts had increased national productivity so much that tax revenue increased overall) and stem-cell research. All the studies found exactly the same thing: if the original dodgy fact fits with your prejudices, a correction only reinforces these even more.

If your goal is to move opinion, this depressing finding suggests that smears work; and what's more, corrections don't challenge them much, because for people who already disagree with you, it only make them disagree even more.

Why Cigarette Packs Matter

Guardian, 12 March 2011

This week our government committed itself to the removal, albeit slowly, of cigarette displays in shops. But plain packaging on cigarettes has been delayed for further consultation.

The Unite union is unimpressed. It represents 6,000 people in tobacco production and distribution, and put out a statement: 'Switching to plain packaging will make it easier to sell illicit and unregulated products, especially to young people'. This 'may increase long-term health problems'. Tory MP Philip Davies said: 'Plain packaging for cigarettes would be gesture politics . . . it would have no basis in evidence.'

Everyone is entitled to their own opinions, but not their own facts. Cigarette packaging has been used for brand building and sales expansion, and that is bad enough; but it has also been used for many decades to sell the crucial lie that cigarettes which are 'light', 'mild', 'silver' and the rest are somehow 'safer'.

This is one of the most important con tricks of all time, because people base real-world decisions on it, even though we have known for several decades that low-tar cigarettes are no safer than normal cigarettes. Manufacturers' gimmicks, like the holes on the filter by your fingers, confuse laboratory smoking machines, but not people. Smokers who switch to lower-tar brands compensate with larger, faster, deeper inhalations, and by smoking more cigarettes. The collected data from a million people shows that those who smoke low-tar and 'ultra-light' cigarettes get lung cancer at the same rate as people who smoke 'normal' cigarettes. They are also, paradoxically, less likely to give up smoking.

So the 'light', 'pale' and 'mild' packaging sells a lie. But do people know this? In data from two population-based surveys, a third of smokers believed incorrectly that 'light' cigarettes reduce health risks, and were less addictive (it's 71 per cent in China). A random telephone digit survey of 2,120 smokers found they believed, on average, that 'ultra lights' convey a 33 per cent reduction in risk. A postal survey of five hundred smokers found a quarter believed 'light' cigarettes are safer. A school-based questionnaire of 267 adolescents found, once again, as you'd expect, that they incorrectly believed 'light' cigarettes to be healthier and less addictive.

Where do all these incorrect beliefs come from? Careful manipulation by the tobacco companies, as you can see for yourself, in their internal documents available for free online. They aimed to deter quitters, and 'mild' products – which were made to seem safer and less addictive – were the perfect vehicle.

But over fifty countries, including the UK, have now banned a few magic words like 'light' and 'mild'. So is that enough? No. A survey of 15,000 people in four countries found that after the ban there was a brief dip in false beliefs in the UK, but by 2005 we bounced back to having the same false beliefs about 'safer-looking' brands as the US.

This is because brand packaging continues to peddle these lies. A street-interception survey from 2009 of three hundred smokers and three hundred non-smokers found that people think packages with 'smooth' and 'silver' in the names are safer, and that cigarettes in packaging with lighter colours, and a picture of a filter, were also safer.

Of course tobacco companies know this. As Philip Morris

said in its internal document 'Marketing New Products in a Restrictive Environment': 'Lower delivery products tend to be featured in blue packs. Indeed, as one moves down the delivery sector, then the closer to white a pack tends to become. This is because white is generally held to convey a clean healthy association.'

If you're in doubt about the impact this branding can have, 'brand imagery' studies show that when participants smoke the exact same cigarettes presented in lighter-coloured packs, or in packs with 'mild' in the name, they rate the smoke as lighter and less harsh, simply through the power of suggestion. These illusory perceptions of mildness, of course, further reinforce the false belief that the cigarettes are healthier.

But these aren't the only reasons why banning a few words from packaging isn't enough. A study on six hundred adolescents, for example, found that plain packages increase the noticeability, recall and credibility of warning labels.

There's no real doubt that the extended, complex, interlocking branding and packaging machinations of cigarette companies play a major role in misleading smokers about the risks. They downplay the harms of smoking, one of the biggest killers in the world, and sadly nothing from Unite – for shame – or some Tory MP will change that.

If you don't care about this evidence, or you think jobs are more important than people killed by cigarettes, or you think libertarian principles are more important than both, then that's a different matter. But if you say the evidence doesn't show evidence of harm from branded packaging, you are simply wrong.

All Bow Before the Mighty Power of the Nocebo Effect

Guardian, 28 November 2009

I'm fine with people wasting their money on sugar pills, but I have higher expectations of government bodies. The Medicines and Healthcare Regulatory Authority has decided to let homeopathy manufacturers make medical claims on their sugar pill bottles, without any evidence of efficacy, and the government funds homeopathy on the NHS. This week the Parliamentary Science and Technology Select Committee looked into the evidence behind these decisions.

There was much cheap comedy – as you'd expect from a government inquiry into an industry based on magical beliefs. But the best moment was when Dr Peter Fisher from the Royal London Homeopathic Hospital (funded by the NHS) explained that homeopathic sugar pills have physical side effects, so they must be powerful. Many raised their eyebrows; but interestingly, the homeopath is correct. People can experience side effects when they receive pills that contain no medicine at all.

A paper published in the journal *Pain* next month looks at this very issue. Its authors found every single placebo-controlled trial ever conducted on a migraine drug, and looked at the side effects reported by the people in the control group, who received a dummy 'placebo' sugar pill instead of the real drug. Not only were these side effects common, they were also similar to the specific side effects you'd have expected if you'd received real drug in the trial: so patients getting placebo instead of anticonvulsants, for example, reported memory difficulties, sleepiness and loss of appetite, which are typical side effects of anticonvulsants; while patients getting placebo instead of painkillers got digestive problems, which themselves are commonly caused by painkillers.

This is nothing new. A study in 2006 sat seventy-five people in front of a rotating drum to make them feel nauseous, and gave them a placebo sugar pill. Twenty-five were told it was a drug that would make the nausea worse: their nausea became worse, and they also exhibited more gastric tachyarrhythmia, the abnormal stomach activity that frequently accompanies nausea.

A paper in 2004 took six hundred patients from three

different specialist drug-allergy clinics, split them into two groups at random, and handed them either the drug that was causing their adverse reactions, or a dummy pill with no ingredients: 27 per cent of the patients experienced side effects such as itching, malaise and headache from the placebo dummy pill.

And a classic paper from 1987 looked at the impact of listing side effects on the consent form which all patients sign before accepting treatment in a randomised trial. This was a large placebo-controlled trial comparing aspirin against placebo, conducted in three different centres. In two of them, the consent form contained a statement outlining various gastrointestinal side effects, and in these centres there was a sixfold increase in the number of people reporting such symptoms, and many more people dropping out of the trial, compared with the one centre that did not list such side effects on the form.

This is the amazing world of the nocebo effect, the evil twin of the placebo effect, where negative expectations can induce unpleasant symptoms in the absence of a physical cause.

Sadly, though, it doesn't help homeopaths: in 2003 Professor Edzard Ernst conducted a systematic review of every single homeopathy trial that reported side effects. This found, in total, fifty episodes of side effects in patients treated with placebo and sixty-three in patients treated with homeopathically diluted remedies, with no statistically significant difference in the rates of side effects between the two groups.

Quacks like to present themselves as holistic, but in reality this research into the placebo effect and the nocebo effect suggests quite the opposite. The world of the homeopath is reductionist, one-dimensional, and built on the power of the pill: it cannot accommodate the fascinating reality of connections between mind and body, which have been revealed in these experiments, and many more. The next time you find yourself trapped, at dinner, next to some bore who's decided they have secret mystical healing powers – while they earnestly explain how their crass efforts at selling sugar pills represent a meaningful political stand against the crimes of big pharma – just think: some lucky person, somewhere in the world, is sat next to a nocebo researcher.

So Brilliantly You've Presented a Really Transgressive Case Through the Mainstream Media

Guardian, 5 December 2009

Here is a mystery. Rom Houben, a Belgian man, was diagnosed as being in a coma for twenty-three years, and he has now made a partial recovery. This has been demonstrated with a series of recently developed brain-scanning techniques (whose predictive value is not entirely known, but they are promising), and he is also opening his eyes. But the story in the media this week goes further than that: it is also claimed that he was conscious all along, but simply unable to move, a well-documented phenomenon called 'locked-in syndrome'. It has been reported as a news story around the world, from the *Sun*, Sky news, the BBC, the *Guardian* (in four separate pieces) and the *Telegraph*, through to CNN, *Der Spiegel*, Australian TV news, and hundreds more.

But there is a problem. Mr Houben has described his horrifying experience of being locked in using something called 'facilitated communication': someone holds his finger; they can sense where his hand wants to go on a screen; and by moving with him, they help him type. If you watch the TV footage, it all happens very fast, too.

What is known about facilitated communication? Many researchers have compared it to ouija boards, which is an attempt, I think, to be fair. Some facilitators may well believe that they are guided by an external force; but there have been several large reviews of research into this technique, and overall, it's not good.

The practice was popular in the 1980s and 1990s, and used mostly in severe autism, so that is where much of the work is found. You might feel this is not entirely applicable to someone with locked-in syndrome; equally, you wouldn't ignore it. A lengthy research review on educational interventions in autism commissioned by the Department for Education and Employment in 1998 found that in FC 'almost all scientifically controlled studies showed that the facilitator was the author of the communication'. This finding was so clear that they concluded further research would be hard to justify.

An academic review in 2001 looked at all the more recent studies, updating two earlier reviews with negative conclusions from 1995, and found that overall, again, the claims made for FC are unsubstantiated.

If you prefer authorities to studies, the National Autistic Society says that five major US professional bodies now formally oppose the use of FC, including the American Academy of Child and Adolescent Psychiatry, the American Speech-Language-Hearing Association, and the American Association on Mental Retardation. The American Psychological Association issued a position paper on FC in 1994 (at the height of its popularity) saying 'studies have repeatedly demonstrated that facilitated communication is not a scientifically valid technique', and calling it 'a controversial and unproved communicative procedure with no scientifically demonstrated support for its efficacy'.

My concern about this is pretty simple. If you watch the video of Mr Houben's facilitated communication in action – and I encourage you to do so – you will see the facilitator looking at the screen and the keyboard, moving Mr Houben's finger at remarkably high speed to type out a message, while both of Mr Houben's eyes are closed, with his head slumped sideways across his chair.

Perhaps this was due to bad video editing. It has also been reported that the facilitated communicator was able to correctly identify objects shown only to Mr Houben in private, although that is a less taxing task than the very rapid one-fingered typing shown on TV. But all of these claims can only be assessed in the context of the overwhelmingly negative research on FC.

Journalists and religious commentators are already writing lengthy moral screeds on the implications of this case for our treatment of people in a coma. That seems premature. Mr Houben's typing may well be genuine, and therefore atypical: nobody can have a meaningful opinion, because newspapers are no place to communicate breakthroughs which are incompatible with large swathes of current knowledge, and based on what seems to be weak and even contradictory evidence.

Now that the amazing case of Mr Houben's facilitated communication has been made the subject of a huge media sensation around the world, and extensive ethical speculation, I think we can all look forward to seeing it formally assessed and presented in an academic paper by his doctor, Professor Steven Laureys of Belgium's Coma Science Group. I've made a note in my diary for this date next year. Just to check.

Prof Laureys never published such a paper. In 2010, after international criticism, he allowed facilitated communication to be tested more rigorously with Mr Houben. It didn't work. This was barely reported.

BAD JOURNALISM

Asking for It

Guardian, 4 July 2009

There's nothing like science for giving that objective, white-coat-flavoured legitimacy to your prejudices, so it must have been a great day for *Telegraph* readers when they came across the headline 'Women who dress provocatively more likely to be raped, claim scientists'. Ah, scientists. 'Women who drink alcohol, wear short skirts and are outgoing are more likely to be raped, claim scientists at the University of Leicester.' Well there you go. Oddly, though, the title of the press release for the same research was 'Promiscuous men more likely to rape'.

Normally we berate journalists for rewriting press releases. Had the *Telegraph* found some news?

I rang Sophia Shaw at the University of Leicester. She was surprised to have been presented as an expert scientist on the pages of the *Daily Telegraph*; because Sophia is an MSc student, and this is her dissertation project. It's also not finished. 'We are intending on getting it published, but my findings are very preliminary.' She was discussing her dissertation at an academic conference, when the British Psychological Society's PR team picked it up, and put out a press release. We will discuss that later.

But first, the science. Shaw spoke to about a hundred men, presented them with various situations around being with a woman, and asked them when they would call it a night, in

329

order to explore men's attitudes towards coercing women into sex. 'I'm very aware that there are limitations to my study. It's self-report data about sensitive issues, so that's got its flaws; participants were answering when sober, and so on.'

But more than that, she told me, every single one of the first four statements made by the *Telegraph* is a flat, factually incorrect misrepresentation of her findings.

Women who drink alcohol, wear short skirts and are outgoing are more likely to be raped? 'We found no evidence that women who are more outgoing are more likely to be raped. This is completely inaccurate. We found no difference whatsoever. The alcohol thing is also completely wrong: if anything, we found that men reported they were willing to go further with women who are completely sober.'

And what about the *Telegraph*'s next claim: its reach for objective distance, and its assertion that it's not just judgemental newspapers, but also *scientists* who claim that provocatively dressed women are more likely to be raped?

'We have found at the minute that people will go slightly further with women who are provocatively dressed, but this result is not statistically significant. Basically, you can't say that's an effect, it could easily be the play of chance. I told the journalist it isn't one of our main findings, you can't say that. It's not significant, which is why we're not reporting it in our main analysis.'

So if the *Telegraph* is throwing blame around with rape, who do we blame for this story, and what do we do about it? On the one hand, Sophia Shaw is not very impressed with the newspaper: 'When I saw the article my heart completely sank, and it made me really angry, given how sensitive this subject is. To be making claims like the *Telegraph* did, in my name, places all the blame on women, which is not what we were doing at all. I just felt really angry about how wrong they'd got this study.' Since I started sniffing around, and Sophia complained, the *Telegraph* has quietly changed the online copy of the article, although there has been no formal correction, and in any case, it remains inaccurate.

But there is a second, less obvious problem. Repeatedly, unpublished work – often of a highly speculative and eye-catching nature – is shepherded into newspapers by the press officers of the British Psychological Society and other organisations. A rash of news coverage and popular speculation ensues, in a situation where nobody can read the academic work. I could only get to the reality of what was measured, and how, by personally tracking down and speaking to an MSc student on the phone about her dissertation. In any situation this would be ridiculous, but in a sensitive area like rape, you'd hope that PR staff could temper their desperate hunger for coverage, and wait for the finished paper.

Jab 'as Deadly as the Cancer'

Guardian, 10 October 2009

Last month I had a debate at the Royal Institution with Lord Drayson, the Science Minister, in which he argued that I was too harsh on British science coverage, which is the best in the world. During this event our chairman (excellently, Simon Mayo) pulled out a health front page from the *Express*, and asked what we thought about it. I said the article might well be accurate, but it's also quite likely to be a work of fantasy. As a serious matter of public health, I would urge people to be extremely sceptical about health information on the front page of the *Express*. Lord Drayson thought this was cynical and unfair. He warmly encouraged us to trust this newspaper.

Here's the latest front-page story from the same paper: 'Jab "as deadly as the cancer".' 'Cervical drug expert hits out as new doubts raised over death of teenager', said the subheading, although no such new doubts were raised in the article. Here is one whole paragraph from that story. Almost every single assertion it makes is false.

The cervical cancer vaccine may be riskier and more deadly than the cancer it is designed to prevent, a leading expert who developed the drug has warned. She also claimed the jab would do nothing to reduce the rates of cervical cancer in the UK. Speaking exclusively to the *Sunday Express*, Dr Diane Harper, who was involved in the clinical trials of the controversial drug Cervarix, said the jab was being 'over-marketed' and parents should be properly warned about the potential side effects.

The story seemed unlikely for three reasons. Firstly, Professor Harper is not a known member of the anti-vaccination community, which is vanishingly small. Secondly, it was on the front page of the *Sunday Express*, which is always cause for concern. Lastly, it was by specialist health journalist Lucy Johnston, whose previous work includes 'Doctor's MMR fears', 'Exclusive: Experts Cast Doubt on Claim for "Wonder" Cancer Jabs', 'Children "Used as Guinea Pigs For Vaccines" ', 'Dangers of MMR Jab "Covered Up" ', 'Teenage Girls Sue Over Cancer Jab', 'Jab Makers Linked to Vaccine Programme', and so many more, including the memorable front-page exclusive: 'Suicides "Linked to Phone Masts"'.

I contacted Professor Harper. To avoid any doubt, I will explain her position on this issue using only her own words: 'I did not say that Cervarix was as deadly as cervical cancer. I did not say that Cervarix could be riskier or more deadly than cervical cancer. I did not say that Cervarix was controversial, I stated that Cervarix is not a "controversial drug". I did not "hit out" – I was contacted by the press for facts. And this was not an exclusive interview.'

Professor Harper did not 'develop Cervarix', as the *Sunday Express* said, but she did work on some important trials of Gardasil, and also Cervarix. 'Gardasil is not a "sister vaccine", as the *Express* said, it is a different compound. I do not know of the side effects of Cervarix as it is not available in the US.' Furthermore, she did not say that Cervarix was being over-marketed. 'I did say that Merck was egregiously overmarketing Gardasil in the US – but Gardasil and Cervarix are not the same vaccines.'

And here is the tragedy. Academics are often independently-minded about the interventions they work on, and Professor Harper – who worked on Gardasil – is critical of Gardasil. More specifically, she is critical of how it is being marketed.

Briefly, her view (which was already published a long time

ago) is that we do not yet know how long the protection from these vaccines will last, and that this will therefore affect the cost-benefit calculations. She is concerned that aggressive advertising aimed directly at the public (which is not permitted in Europe, with good reason) may lead people to falsely believe they are immune to HPV, which causes cervical cancer, and so neglect other precautions. Lastly, she suspects from modelling data that, for the specific and restricted group of women who are punctilious about attending every single one of their cervical-cancer screening appointments, vaccination may have little impact on their risk of death from cancer; but she also says that even this group will still benefit from the reduction in reproductive problems caused by treating precancerous changes in cervical cells, and from avoiding the unpleasantness of screening and treatment.

The article has now disappeared from the *Express* website, and Professor Harper has complained to the PCC. 'I fully support the HPV vaccines,' she says. 'I believe that in general they are safe in most women. I told the *Express* all of this.'

Her criticisms of some aspects of cervical-cancer vaccination are nuanced and valuable: but they don't fit into the black-and-white hysteria of British news media. It would be nice if we could have a serious public discussion about the relative risks and merits of different vaccine options. Sadly, with this kind of ugly reporting, scientists around the world may learn that such a discussion is not currently possible in the UK media. That is the greatest tragedy.

Health Warning: Exercise Makes You Fat

Guardian, 29 August 2009

Why would you listen to a government health message, or your GP practice nurse, when the *Sunday Telegraph* has much more exciting news? 'Health Warning: Exercise Makes You Fat' is the kind of full-width headline you want to see across a broadsheet page: it's affirmative, it's reassuring, and it gives you clear permission to sit on your arse all day. 'Re-programming body fat is the key to weight loss, not working out.' Praise be. 'Is it possible that all that exercise is doing nothing to make us slimmer?' Please, let the answer be yes.

The *Telegraph* produced three lines of research for this claim. Firstly, more people are spending more money on more exercise than before, but there is also more obesity around in the UK than before: explain *that* with your science. Then there was some speculative laboratory research about interfering with brown fat in animal models, using stem cells and things: interesting to read, but very far from the headline claim.

To properly examine whether exercise really will make you fat, the paper described two trials.

The first one, I can tell you right now, is cherry-picked. The Cochrane Library is a non-profit collaboration of academics who produce unbiased, systematic reviews of the medical literature, and they have a systematic review of all the forty-three trials that have been done on exercise for weight loss. This produces clear evidence that exercise is beneficial, albeit more modestly than you'd hope. 'Exercise plus diet' was compared with 'diet alone' in fourteen trials: both groups lost weight, but 1.1 kg more in the exercise group. High-intensity exercise was compared with low-intensity in four trials: high-intensity exercise came out better in all of them, with extra weight loss of 1.5 kg. There are also improvements in blood pressure, cholesterol, blood sugars, sense of well-being, and so on.

The *Telegraph* quoted one trial from Dr Timothy Church of Louisiana University, which compared three different levels of exercise with a personal trainer in overweight people. There were no significant differences between the weight lost in any of the groups, including the 'control' group, who were not given a personal trainer at all. So while it is true that exercise did indeed have no benefit, in the one single trial the *Telegraph* quoted, it has also ignored the vast, overwhelming majority of published literature examining the same question. Dr Church speculates that the explanation for his finding is that people who exercised more also ate more. Dr Church is speculating in order to explain the odd results of his one single trial. I think the most helpful suggestion he could make here would be: 'Our unusual results are probably a fluke, because almost all other trials using similar methods found a completely different answer.'

Then there is the *Telegraph*'s second trial. 'Another study due to be published next month in the *Journal of Public Health Nutrition* by researchers at the University of Leeds draws similar conclusions. Professor John Blundell and his colleagues found that people asked to do supervised exercise to lose weight also increased the amount they ate and reduced their intake of fruit and vegetables.'

I have this trial in front of me. It's simply not true that participants increased their food intake. A tiny proportion did (15 per cent), but that's hardly an issue, because what the *Telegraph* doesn't tell you – bizarrely – is that overall, participants doing supervised exercise in this trial lost more weight. Much more weight: 3.2 kg more, on average, over just twelve weeks.

Prof Blundell says: 'The *Telegraph* article was a complete distortion of the facts of our investigation, which showed that exercise is very effective for weight loss. They completely reversed the outcome of our study.'

You might well view my work as pointless: like Sisyphus in an anorak, fighting my way up a greasy waterslide, defeated by the torrent of sewage, desperately trying to scratch one grumpy correction into yesterday's chip wrapper. But journalism like this is a genuine public health problem. Research has repeatedly shown that people change their health behaviour in response to what they read in the media, and just this month, the World Cancer Research Fund commissioned a survey from YouGov: a proper survey, in a representative sample, from a reputable data collector, where anyone is allowed to see the questions and the results.

Half of all respondents said they thought scientists and doctors were constantly changing their minds about healthy-living advice, although in reality healthy-living advice hasn't changed at all for at least a decade (don't smoke, do some exercise, eat more fruit and veg). And a quarter of all respondents said that because scientists keep changing their minds, you might as well eat whatever you want, because it won't make any difference anyway.

Have another pastry and put the telly on.

The Caveat in Paragraph Number 19

Guardian, 16 October 2010

You will be familiar with the *Daily Mail*'s ongoing project to divide all the inanimate objects in the world into the ones that either cause or prevent cancer. It's hardly worth documenting the individual cases any more: you can appreciate the phenomenon in bulk, through websites like the *Daily Mail* Oncological Ontology Project and Kill Or Cure, with its alphabetised list: from almonds, apples and artificial light; through horseradish, hot drinks and housework; to wasabi, water, watercress, and more.

But occasionally one story pops up to illustrate a wider issue, and 'Strict diet two days a week "cuts risk of breast cancer by 40 per cent" ' is a good example. It goes on: 'A strict diet for two days a week consisting solely of vegetables, fruit, milk and a mug of Bovril could prevent breast cancer, scientists say.'

Now, if you read the academic paper which this news article is describing, from the *International Journal of Obesity*: it's not a study of breast cancer, and it does not find that the risk of cancer is reduced by 40 per cent. The press release wasn't exactly a masterpiece of clarity either, but in any case, the study doesn't even measure breast cancer as an outcome, at all.

If I was to leave it there, the journalist would correctly complain: because after all the grand and misleading claims, firstly, briefly, in the body of the piece, it does mention that the outcome is not cancer, but some hormones related to cancer (with no explanation of how tenuous that relationship is). Then, finally, at the very bottom of the piece, comes the reality. Although it's not spoken in the authoritative third person of the paper itself, it's there, in a quote, at paragraph number 19:

> But Dr Julie Sharp, senior science information manager at Cancer Research UK, said: 'This study is not about breast cancer, it's a study showing how different diet patterns affect weight loss and it's misleading to draw any conclusions about breast cancer from this research.'

The late caveat, torpedoing the central premise of a news piece, is a common strategy in many newspapers. But what use is this information, at the end of a long article, in paragraph number 19?

The way people read newspapers has been studied widely, using eye-tracking technology. It's through this that we discover, for example, that when presented with a full-length photograph of a man, men are more likely to look at the penis area than women.

Most of this research is more interested in adverts than news, because research in all fields is driven by money (top left of the page is best, apparently): but there is plenty of other useful stuff, much of it by the Poynter Institute.

They did an early study in 1990, with predictable findings: photos attract attention; eyes travel from the dominant photo to the biggest headline, then teasers, and finally text; text is read the least, headlines the most; and so on.

But their most recent project was far bigger: they took a representative sample of 582 people from four cities in the US, and invited them to read a newspaper and a website as they normally would, wearing the eye-tracking equipment, over five days in 2006, for fifteen minutes each. This yielded a dataset of more than 102,000 eye stops.

Here's what they found: once a story gets longer than eleven paragraphs, on average, your readers will read only half. A tiny minority will make it to paragraph number 19, where, on this occasion, a fraction of the readers of the *Daily Mail* would have discovered that the central premise of the news story – that a new trial had found a 40 per cent reduction in cancer through intermittent dieting – was false.

Caveats in paragraph 19 are standard practice for stories with outlandish health claims. Like nipple tassels in 1950s burlesque, they're a way to keep it legal, but titillating; and in many cases, when the late rebuttal comes from an authority figure – calling for calm in patrician tones – it can feel as if it's only even there to accentuate the excitement. But if your interest is informing a reader, they are plainly misleading.

Why Don't Journalists Link to Primary Sources?

Guardian, 19 March 2011

Why don't journalists link to primary sources? Whether it's a press release, an academic journal article, a formal report, or even the full transcript of an interview, the primary source contains more information for interested readers: it shows your working, and it allows people to check whether what you wrote is true. Here are three short stories.

This week the *Telegraph* ran the headline 'Wind farms blamed for stranding of whales'. 'Offshore wind farms are one of the main reasons why whales strand themselves on beaches, according to scientists studying the problem,' it continued. Baroness Warsi even cited the story on BBC *Question Time* this week, arguing against wind farms.

But anyone who read the open-access academic paper in *PLoS One*, titled 'Beaked Whales Respond to Simulated and Actual Navy Sonar', would see that the study looked at sonar, and didn't mention wind farms at all. At best, the *Telegraph* story was a massive exaggeration of one brief, contextual aside about general levels of manmade sound in the ocean, made by one author at the end of the press release (titled 'Whales Scared by Sonars'). This release didn't mention wind farms, and it didn't say they were 'one of the main reasons why whales strand themselves on beaches'. Anyone reading the press release could see that the study was about naval sonar.

This *Telegraph* article (now deleted, with a miserly correction) was a distortion, perhaps driven by the paper's odd editorial line about the environment. But there is a bigger fish here: if we had a culture of linking to primary sources – if they were always just a click away – then shame alone would probably have stopped it going online. Outright misrepresentations are only worth risking in an environment where the reader is routinely deprived of information.

Sometimes the examples are sillier. Professor Anna Ahn published a paper recently, showing that people with shorter heels have larger calves. For the *Telegraph* this became 'Why stilettos are the secret to shapely legs', for the *Mail* 'Stilettos give women shapelier legs than flats', for the *Express* 'Stilettos tone up your legs'.

But anybody who read even the press release would immediately see that this study had nothing to do with shoes. It wasn't about shoe-heel height: it looked at anatomical heel length, the distance from the back of your ankle joint to the insertion of the Achilles tendon. The participants were all barefoot, and the paper was just a nerdy insight into the engineering of a human body: if you have a shorter lever at the back of your foot, you need a bigger muscle in your calf. Once again, this story was a concoction by journalists. But more than that, no sane journalist could possibly have risked writing the story about stilettos, if there was a culture and tradition of linking to the academic paper, or even the press release: they'd have looked like idiots, and fantasists, to anyone who bothered to click.

Lastly, on Wednesday the *Daily Mail* ran with the scare headline 'Swimming too Often in Chlorinated Water "Could Increase Risk of Developing Bladder Cancer", Say Scientists'. There's hardly any point documenting the errors in *Daily Mail* health stories any more, but if you read the original paper, or even the press release, again, anyone can see that bladder cancer wasn't measured, and the *Mail*'s story was a simple distortion. It's worth mentioning that these press releases were fairly readable pieces of popular science in themselves.

Of course, this is a problem that occurs well beyond science. Over and again, you read comment pieces that purport to be responding to an earlier piece, but distort the earlier arguments, or miss out the most important ones: they count on it being inconvenient for you to check. It's also an interesting difference between different forms of media: most bloggers have no institutional credibility, so they must build it, by linking transparently, and allowing you to easily double-check their work.

But more than anything, because linking sources is such an easy thing to do, and the motivations for avoiding links are so dubious, I've detected myself using a new rule of thumb: if people don't link to primary sources, I don't trust them, and I don't read them.

A Fishy Friend, and His Friends

Guardian, 5 June 2010

'Fish oil helps schoolchildren to concentrate' was the headline in the *Observer*. The omega-3 fish-oil pill issue has dragged on for almost a decade now: the entire British news media repeatedly claim that trials show it improves school performance and behaviour in mainstream children, but no such trial has ever been published. There is something very attractive about the idea that solutions to complex problems in education can be found in a pill.

So, have things changed? The *Observer*'s health correspondent, Denis Campbell, is on the case, and it sounds as if they have. 'Boys aged eight to 11 who were given doses once or twice a day of docosahexaenoic acid, an essential fatty acid known as DHA, showed big improvements in their performance during tasks involving attention.' Great. 'The researchers gave 33 US schoolboys 400mg or 1,200mg doses of DHA or a placebo every day for eight weeks. Those who had received the high doses did much better in mental tasks involving mathematical challenges.' Brilliant news.

Is it true? After some effort, I have tracked down the academic paper. This was not a trial of whether fish-oil pills improve children's performance: it was a brain-imaging study. They took thirty-three kids, divided them into three groups (of ten, ten and thirteen) and then gave them either no omega-3, a small dose, or a big dose. Then the children performed some attention tasks in a brain scanner, to see if bits of their brains lit up differently.

Why am I saying 'omega-3'? Because it wasn't a study of fish oil, as the *Observer* says: it was a study of omega-3 fatty acids derived from algae. But that's small print.

If this had been a trial to detect whether omega-3 improves performance, it would be laughably small: about ten people in each treatment group. While small studies aren't entirely useless, as amateurs often claim, you do have a very small number of observations to work from, so your study is much more prone to error from the simple play of chance. A study with thirty-three children, like this one, could conceivably detect an effect; but only if the fish oil caused a gigantic and unambiguous improvement in all the children who got it, and none of the children on placebo improved.

This paper showed no difference in performance at all. Since it was a brain-imaging study, not a trial, the results of the children's actual performance on the attention task are reported only in passing, in a single paragraph, but the researchers are clear: 'There were no significant group differences in percentage correct, commission errors, discriminability, or reaction time.'

So this is all looking pretty wrong. Are we even talking about the same academic paper? I've been trying to get mainstream media to link to original academic papers when they write about them, at least online, with some limited success on the BBC website. I asked Denis Campbell which academic paper he was referring to, but he declined to answer, and passed me on to Stephen Pritchard, the Readers' Editor for the *Observer*, who answered a couple of days later to say that he didn't understand why he was being involved. Eventually Denis confirmed, but through Stephen Pritchard, that it was indeed the paper I had found, from the April edition of the *American Journal of Clinical Nutrition*.

If we are very generous, is it informative, in any sense, that a brain area lights up differently in a scanner after some pills? Intellectually, it may be. But doctors get very accustomed to drug company sales reps and enthusiastic researchers who approach them with an exciting *theoretical* reason why one treatment should be better than another: maybe their intervention works selectively on only one kind of receptor molecule, for example, so it should therefore have fewer side effects. Similarly, drug reps and researchers will often announce that their intervention has an effect on some kind of elaborate laboratory measure: maybe a molecule in the blood goes up in concentration, or down, in a way that suggests the intervention might be effective.

This is all very well. But it's not the same as showing that something really does work, back here in the real world, and medicine is overflowing with unfulfilled promises from early theoretical research. This stuff is interesting: but it's not even in the same ballpark as showing that something works.

Oddly enough, though, someone really has finally conducted a proper trial of fish-oil pills in mainstream children to see if they work. It's the trial that journalists have long been waiting for: a well-conducted, randomised, double-blind, placebo-controlled trial, in 450 children aged eight to ten from a mainstream school population. It was published in full this year, and it found no improvement. Show me the news headlines about that paper.

Meanwhile, Euromonitor estimates global sales for fish-oil pills at $2 billion, having doubled in five years, with sales projected to reach $2.5 billion by 2012, and they are now the single best-selling product in the UK food-supplement market. This has only been possible with the kind assistance of the British media, and their eagerness to write stories about the magic intelligence pill.

The week after this piece appeared, the Independent's *health correspondent wrote an angry column, explaining that health correspondents can't be expected to check facts. In the interests of balance, his piece is reproduced below in full.*

Jeremy Laurance: Dr Goldacre Doesn't Make Everything Better

Is Ben Goldacre, the celebrated author of *Bad Science* and scourge of health journalists everywhere, losing it? So accustomed has he become to swinging his fists at the media when they get a science story wrong, I fear he may one day go nuclear and take out three rows of medical correspondents with a single lungful of biting sarcasm.

He was at it again in Saturday's *Guardian*, pistol-whipping his *Guardian* and *Observer* colleague, health correspondent Denis Campbell, over a report he wrote about fish oil and its supposed role in improving children's intelligence.

Campbell had reported claims made at a press conference that fish oil improved mental performance in children taking supplements. His crime, however, was to fail to check the claims against the academic paper on which they were based. That showed that the fish oil 'enhanced the function of those brain regions that are involved in paying attention', as revealed by a brain scanner.

Not quite the same as 'improving their performance', as Goldacre rightly pointed out. Indeed the paper revealed that there had been no improvement in the children's performance. Time, then, for Goldacre to deliver his customary knee-capping. He did so because Campbell declined to help him with his inquiries. Small wonder, given it is the second occasion the hapless Campbell has found himself in Goldacre's sights.

One doesn't know whether to laugh or cry at the *Guardian*'s eagerness to wash its dirty linen in public. It is undeniably magnificent, but – in my view – no way to run a newspaper. I wonder at the psychiatrist's bills. What does it tell us about health and science reporting? First, most disinterested observers think standards are pretty high (a report by the Department of Business last January said it was in 'rude health'). Second, reporters are messengers – their job is to tell, as accurately as they can, what has been said, with the benefit of such insight as their experience allows them to bring, not to second guess whether what is said is right. But third, reporters are also under pressure. Newspaper sales are declining, staff have been cut, demands are increasing.

Goldacre is right to highlight the fact that there is too much 'churnalism' – reporters turning out copy direct from press conferences and releases, without checking, to feed the insatiable news machine. This ought to be stopped. But no one, so far, has come up with a commercially realistic idea of how to stop it.

In the meantime, while raging rightly at the scientific illiteracy of the media, he might reflect when naming young, eager reporters starting out on their careers that most don't enjoy, as he does, the luxury of time, bloggers willing and able to do his spadework for him (one pointed out the flaws in Campbell's report on the *Guardian* website five days before Goldacre's column appeared) and membership of a profession (medicine) with guaranteed job security, a comfortable salary and gold-plated pension. If only.

Make of that what you will. Jeremy Laurance mentioned that this is the second time I've written about a piece by Denis Campbell. Below is the first, a front-page splash by Denis in the Observer.

MMR: The Scare Stories Are Back

British Medical Journal, 18 July 2007

It was inevitable that the media would re-ignite the MMR (measles, mumps, rubella) autism scare during Andrew Wakefield's General Medical Council hearing. In the past two weeks, however, a front-page splash in the *Observer* has drawn widespread attention: the newspaper effectively claimed to know the views of named academics better than those academics themselves, and to know the results of research better than the people who did it. Smelling a rat – as one might – for once, I decided to pursue every detail.

The *Observer*'s story made three key points: that new research had found an increase in the prevalence of autism, to one in fifty-eight; that the lead academic on this study was so concerned he suggested raising the finding with public health officials; and that two 'leading researchers' on the team believed that the rise was due to the MMR vaccine. By the time the week was out, this story had been recycled in several other national newspapers, and the one in fifty-eight figure had even been faithfully reproduced in a *BMJ* news article.

On every one of these three key points the *Observer* story was simply wrong.

The newspaper claimed that an 'unpublished' study from the Autism Research Centre in Cambridge had found a prevalence for autism of one in fifty-eight. I contacted the centre: the study that the *Observer* reported is not finished, and not published. The data have been collected, but they have not been analysed.

Unpublished data is a recurring theme in MMR scares, and it is the antithesis of what science is about: transparency, where anyone can read the methods and results, appraise the study, decide for themselves whether they would draw the same conclusions as the authors, and even replicate, if they wish.

The details of this study illustrate just how important this transparency is. It was specifically designed to look at how different methods of assessing prevalence affected the final figure. One of the results from the early analyses is 'one in fifty-eight'. The other figures were less dramatic, and similar to current estimates. In fact the *Observer* now admits it knew of these figures, and that these should have been included in the article. It seems it simply cherry-picked the single most extreme number – from an incomplete analysis – and made it a front-page splash story.

And why was that one figure so high anyway? The answer is simple. If you cast your net as widely as possible, and use screening tools, and many other methods of assessment, and combine them all, then inevitably you will find a higher prevalence than if – for example – you simply trawl through local school records and count your cases of autism from there.

This is not advanced epidemiology, impenetrable to journalists – this is basic common sense. It would not mean that there is a rise in autism over time, compared with previous prevalence estimates, but merely that you had found a way of assessing prevalence that gave a higher figure. More than that, of course, when you start doing a large-scale prevalence study, you run into all kinds of interesting new methodological considerations: Is my screening tool suitable for use in a mainstream school environment? How does its positive predictive value change in a different population with a different baseline rate? And so on.

These are fascinating questions, and for that reason statisticians and epidemiologists were invented. As Professor Simon Baron-Cohen, lead author on the study, says: 'This paper has been sitting around for a year and a half specifically because we've brought in a new expert on epidemiology and statistics, who needs to get to grips with this new dataset, and the numbers are changing. If we'd thought the figures were final in 2005, then we'd have submitted the paper then.'

The *Observer*, however, is unrepentant: it has the 'final report'. And what is this document? I can't get the paper to show it to me (and what kind of a claim about scientific evidence involves secret data?), but grant-giving agencies expect a report every quarter, right through to the end of the grant, and it seems likely that what the *Observer* has is simply the last of those: 'That might have been titled "final report"', said Professor Baron-Cohen. 'It just means the funding ended, it's the final quarterly report to the funders. But the research is still ongoing. We are still analysing.'

But these are just nerdy methodological questions about prevalence (if you skip to the end, there is some quite good swearing). How did the *Observer* manage to crowbar MMR into this story? Firstly, it cranked up the anxiety. According to the newspaper, Baron-Cohen 'was so concerned by the one in fifty-eight figure that last year he proposed informing public health officials in the county'.

But Professor Baron-Cohen is clear: he did no such thing, and this was simply scaremongering. I put this to the *Observer*, which said it had an email in which Baron-Cohen did as the paper claimed. *Observer* staff gave me the date. I went back to the professor, who went through his emails. We believe that I too now have the email to which the *Observer* refers. It is one sentence long, and it is Professor Baron-Cohen asking if he can share his and the other researchers' progress with a clinical colleague in the next-door office. This dramatic smoking gun reads: 'can i share this with ayla and with the committee planning services for AS [autism services] in cambridgeshire if they treat it as strictly confidential?'

Professor Baron-Cohen told me, 'That's not saying I'm concerned, or that we should notify anybody; these are just the people who run the local clinic, who I share a corridor with, who said they were interested to hear how it was going so far. They are not public health officials, and it's not alarmist, it's not voicing concern, it's simply saying: "Am I allowed to share a paper with a colleague in the next-door office?" It seems very important to me that we discuss clinical research with clinical colleagues, and I only stressed confidentiality because the paper was not yet final.'

But what about the meat? The *Observer* claims that 'two of the academics, leaders in their field, privately believe that the surprisingly high figure [one in fifty-eight] may be linked to the use of the controversial MMR vaccine'. This point is repeatedly reiterated, with a couple of other scientists disagreeing so as to create that familiar, unnerving, illusory equipoise of scientific opinion that has fuelled the MMR scare in the media for almost a decade now.

These ridiculous baby bibs are available from www.badscience.net, and they are an excellent way to start interesting conversations with other parents. (Nobody ever buys them.)

But in fact the two 'leading experts' concerned about MMR were not professors, or fellows, or lecturers: they were research associates. I rang both, and both were very clear that they wouldn't really describe themselves as leading experts. One is Fiona Scott, a psychologist and very competent researcher at Cambridge. She said to me: 'I absolutely do not think that the rise in autism is related to MMR.' And: 'My own daughter is getting vaccinated with the MMR jab on July 17.'

She also says, astonishingly, that the *Observer* never even spoke to her before incorrectly reporting that she has a privately held view that MMR might be partly to blame for autism. I say 'reporting', but in some ways it's more like an accusation. Dr Scott was horrified. She simply does not believe that MMR has caused a rise in autism.

And yet the *Observer*'s 'Readers' Editor' column one whole week later (15 July), when the paper half-heartedly addressed some (and I mean some) of the criticisms of its piece, reinforced the idea that she holds this view. It's like a repeating nightmare. They say they know she does, because of a report she wrote, from 2003.

I trudged back to Scott. Firstly, she tells me, this was a legal report pertaining to a specific group of disabled children, submitted four years ago. Secondly, her view has been, and is, that if MMR has a causative role in symptoms which fit into a diagnostic category of autism (which is after all very broad, and can easily subsume a lot of children with learning difficulties and organic injury), then the numbers are so small that they are not in excess of what people already routinely expect in side effects from vaccines: again, she does not think MMR causes a rise in autism prevalence.

But lastly, even if she ever had, in 2003, in one report, made such a suggestion, one is entitled to change one's view in over four years of working and researching, and in the face of new evidence and experience; and indeed that would be admirable.

If you're saying someone holds a view, right now, and you haven't even asked them, and they loudly say they don't, then it doesn't really matter what you think you've read in a court report from half a decade ago: you're on very shaky territory.

But the paper still has not contacted Fiona Scott: apparently the *Observer* knows the opinions of this woman better than she knows her own mind, despite her public protestations. In fact the only voice Dr Scott could find (while the *Observer* continued to describe her as a 'dissenter') was in the online comments underneath the Readers' Editor piece, where she posted an impassioned and rather desperate message. I shall reproduce it almost in full, because she deserves the space the print *Observer* has repeatedly denied her:

> I feel, given that I was one of the two 'leaders in the field' (flattering, but rather an exaggeration) reported as linking MMR to the rise in autism, that I should quite clearly and firmly point out that I was never contacted by and had no communication whatsoever with the reporter who wrote the infamous *Observer* article. It is somewhat amazing that my 'private beliefs' can be presented without actually asking me what they are. What appeared in the article was a flagrant misrepresentation of my opinions – unsurprising given that they were published without my being spoken to.
>
> It is outrageous that the article states that I link rising prevalence figures to use of the MMR. I have never held this opinion. I do not think the MMR jab 'might be partly to blame'. As for it being a factor in 'a small number of children', had the journalist checked with me it would have been clear that my view is in line with Vivienne Parry of the JCVI [Joint Committee on Vaccination and Immunisation]. The 'small number' was misrepresented by being linked inappropriately and inaccurately with 'rise in prevalence', leading readers to arguably infer that it is in fact NOT a small number!
>
> I wholeheartedly agree with Prof Baron-Cohen, and many

of the posts and responses received to date, that the article was irresponsible and misleading. Furthermore I reiterate that it was inappropriate in including views and comments attributed to me and presented as if I had input into the article when I had not (and still have not) ever been contacted by the journalist in question. I am taking the matter under advisement.

The story takes one final, bizarre turn. The last 'leading expert' is Carol Stott. She does believe that MMR causes autism (at last!). However, she is no longer even a 'research associate' at the Autism Research Centre. Carol Stott works in Dr Andrew Wakefield's private autism clinic in America, and she was also an adviser to the legal team that failed in seeking compensation for parents who believed that MMR caused their children's autism. She was paid £100,000 of public money for her services. She says her objectivity was not affected by the sum, but to me this is not the issue: it seems an astonishing pair of facts for the *Observer* to leave out of its original article. It is still not satisfactory for this kind of information to appear only one week later in the Readers' Editor's column. 'Conflict of interest' is a situation, not a behaviour.

And were her views private, and unknown? No. Stott is so committed to the cause against MMR that when the investigative journalist Brian Deer exposed the legal payouts in 2004, although she had no prior contact with him, she spontaneously fired off a long series of sweary emails titled 'game on': 'Try me, shit head . . . Beleive [sic] me, you will lose . . . so go fuck yourself. Got it yet shit head. Try me . . . Twathead . . . waiting . . . oh yes . . . Stick that where it feels good. shit head . . . well, ur a bit slow on the uptake . . . Give it time I s'pose. twat.' And so on.*

---

* Although it wasn't adequately recognised at the time, I feel I should take some credit for getting 'shit head', 'fuck yourself', 'twathead' and 'twat' into a top-ten academic journal. The gauntlet is down.

I rang her with some trepidation, but when I got through I instantly and genuinely warmed to her. She regrets that many people have fallen into entrenched positions over MMR on both sides, including herself. But she's not a leading expert (as she herself agrees); she's hardly a senior Cambridge academic suddenly expressing a fresh and private concern (her views are perfectly public); and even she is very clear that this new research tells us nothing whatsoever about MMR causing autism.

Outside of the details, there is a wider story here. The media have diligently avoided writing anything on the negative findings in autism research. Instead they have chosen repeatedly to concoct huge stories from the 'concerns' of 'experts' and research that is unpublished and inaccessible, locked away in a box, where they can say what they like about it. They can even refuse – as the *Observer* has with my approaches – to actually show their evidence; and that is the absolute polar opposite of what science and evidence-based opinions are about.

On this occasion I was able to go to the source, and debunk the *Observer*'s claims: often, though, these 'new unpublished research' stories are concocted with the complicity of researchers from the anti-vaccine movement, and there is no such luck.

Whatever one might think about Andrew Wakefield, he was just one man; the MMR autism scare has been driven for a decade now by a media that over-emphasises marginal views, misrepresenting and cherry-picking research data to suit its cause. As the *Observer* scandal makes clear, there is no sign that this will stop.

Prevention Is Better than Cure When It Comes to Health Scares

Financial Times, 2 May 2013

Swansea in Wales is experiencing a serious measles outbreak, with 1,000 cases and one suspected death so far. The outbreak is the result of an unfounded fear that the combined measles, mumps and rubella (MMR) vaccine caused autism. This is just one among dozens of vaccine scares, and more will inevitably come if no action is taken.

Antivaccination campaigners have existed for as long as vaccination itself, using the same arguments over three centuries. If these scares were driven by evidence, we would expect them to appear everywhere at once. But vaccine panics respect local cultural boundaries, because they are cultural phenomena, driven by social and political factors.

In France, for example, there was a significant scare during the 1990s that a hepatitis B vaccine caused multiple sclerosis. And yet this was barely heard of outside France, even though the vaccine is offered universally in most developed nations. The UK's MMR scare was most active between 1998 and 2003, creating 1,200 news articles in one year at its peak, and yet this was barely noticed outside the UK.

The US, meanwhile, had its own scare about vaccines and autism, centred on a preservative called thiomersal, which peaked half a decade after the UK MMR–autism scare. The US scare, in turn, received little coverage in mainstream European media.

Meanwhile in 2002, the World Health Organization was on target to eradicate polio from the face of the globe, when a vaccine scare arose across Kano province in northern Nigeria, with imams claiming that the vaccines were part of a plot by drug companies – and the US – to make Muslims sterile. Polio spread to neighbouring nations as well as Sudan, Yemen and Indonesia. But the scare also spread, often along religious channels, reaching Pakistan, where it contributed to the murder of seven vaccination workers last year in Karachi.

Local politics are a recurring theme in triggering these outbreaks. Suspicion around polio vaccinators in Pakistan was heightened by the US Central Intelligence Agency using the vaccination programme as a cover to find Osama bin Laden. The Kano scare emerged in the same area of Nigeria where Pfizer's Trovan drug trial – associated by many with John le Carré's novel *The Constant Gardener* – was blamed for eleven deaths just a few years earlier, and just as the case came to court in the US.

The peak of the UK MMR scare occurred in 2001, a full three years after the publication of the research – now forcibly retracted – on which it was based. It was driven by media speculation over the reluctance of Tony Blair, then prime minister, to disclose whether his eighteen-month-old son had had the vaccine.

But while the triggers may be political, scares land on fertile ground, and nothing has been done to address these background factors. In the UK, there is clear evidence to implicate irresponsible journalism. The *South Wales Evening Post* covers the area of the Swansea outbreak, and ran a notably aggressive anti-MMR campaign. Published academic research has shown that the MMR vaccine uptake in this one newspaper's distribution area dropped 13 per cent, while coverage in the rest of Wales fell only 2 per cent. But many journalists remain reluctant to accept responsibility, while public health authorities are keener to promote positive messages than to confront misinformation.

Our systems in science to manage misleading content are also flawed, or even non-existent. Andrew Wakefield, author of the flawed MMR research, was only struck off the medical register after a decade's delay. Retracting the paper took just as long. The research misconduct in his work was ultimately exposed by Brian Deer, an investigative journalist, while the institutions and academic journals involved were reluctant to accept responsibility for investigating at all. Again, little has changed here over the past decade.

It is right to deploy resources on catch-up programmes for MMR and other missed vaccines when outbreaks occur. But prevention is better than cure, and history teaches us that fresh vaccine scares will always emerge. Eradication is a distant dream, but better education, early media monitoring and improved accountability through scientific institutions may help.

Dodgy Academic PR

Guardian, 30 May 2009

Obviously we distrust the media on science: they rewrite commercial press releases from dodgy organisations as if they were health news, they lionise mavericks with poor evidence, and worse. But journalists will often say: What about those scientists with their press releases? Surely we should do something about them, running about, confusing us with their wild ideas?

You may think that a journalist should be able to do more than simply read, and then rewrite, a press release; but we must accept that these are troubled times.

So, in this imperfect world, it would be useful to know what's in academic press releases; we are entitled, after all, to expect a very high standard of factual accuracy from academics. Sadly, a new paper in the *Annals of Internal Medicine* this month shows that we have been failed.

Researchers at Dartmouth Medical School in New Hampshire took one year's worth of press releases from twenty medical research centres. This was a mixture of the most eminent universities and the most humble, as measured by their US News & World Report ranking. Those centres each put out around one press release a week. Two hundred were selected at random and analysed in detail.

Half covered research done in humans, and as an early clue to their quality, 23 per cent didn't bother to mention the number of participants – it's hard to imagine anything more basic – and 34 per cent failed to quantify their results. What kinds of study were covered? In medical research we talk about the 'hierarchies of evidence', ranked by quality and type. Systematic reviews of randomised trials are considered to be the most reliable: because they ensure that your conclusions are based on all of the infor- mation, rather than just some of it; and because randomised trials – when conducted properly – are the least vulnerable to bias, and so they are the 'most fair tests'. After these, there are observational studies: these are much more prone to bias, and produce findings which might just reflect correlation instead of causation ('People who choose to eat vegetables live longer') but they are generally cheaper to do. Then there are individual case reports. And then, finally, because medical academics like to think they're funny, right at the bottom of the hierarchy you will find something called 'expert opinion'.

In the Dartmouth study, among the press releases covering human research, only 17 per cent promoted the studies with the strongest designs, either randomised trials or meta-analyses. Forty per cent were on the most limited studies: ones without a control group, small samples of fewer than thirty participants, studies looking at 'surrogate primary outcomes' (which might measure a blood-cholesterol level, for example, rather than something concrete like a heart attack), and so on.

That's not necessarily a problem. Research is always a matter of compromise over what is practical, or affordable. It would be nice to randomise every single patient, everywhere, whenever there is any uncertainty over which is the best treatment for their condition, and perfectly follow their progress, but that would be quite a piece of work. It would be nice to randomise everyone in the country to different lifestyle choices at birth, to see which had the most significant impact on their health, so that in seventy years' time we would have a comprehensive story on the best way to live; but it's hard enough to get people recruited and cooperating in a brief three-week study, let alone lifelong change; and in any case, who wants an answer in the year 2079?

So people conduct imperfect research, knowing that it is the best we can do with the resources available, knowing that the results must be interpreted with caution and caveats. This isn't 'bad science', because the studies themselves are – we assume – well conducted, and faithfully described in their publications. The errors come at the level of interpretation, where people fail to acknowledge the limitations of the evidence.

That failure is a crime, but is it limited to quacks and hacks? No, and that is the key finding of this new paper. Fifty-eight per cent – more than half – of all press releases from this representative sample of academic institutions lacked the relevant cautions and caveats about the methods used and the results reported. I would like journalists to be experts in their field – and I don't think they could be bluffed as easily by a politician or a sports personality as they are by a science press release – but make no mistake, this is a war on all fronts.

Suicide

Guardian, 28 March 2009

This week, in my crescendo-ing tirade against journalism, we shall review the evidence that journalists kill.

The suicide of Sylvia Plath's son has filled the news. The media obsessed – understandably – over genetics, when mental illness is probably the single biggest risk factor, but the coverage has been universally thoughtful, considerate, informed and responsible. This is not always the case, as we shall see. But before we get there, one important cause of suicide seems to have been missed.

In *The Sorrows of Young Werther* by Goethe, the hero shoots himself because his love is unattainable. The novel was banned after young men throughout Europe were reported to be dressing like Werther, copying his affectations, and taking their own lives in the same style.

But a myth about a book is not enough: you need research. And it has been shown repeatedly that suicide increases in the month after a front-page suicide story. There is also evidence that the effect is bigger for famous people and gruesome attempts. You may want to remember that fact for later.

Details matter, as ever. Overdoses increased by 17 per cent in the week after a prominent overdose on *Casualty* (watched by 22 per cent of the British population at the time), and paracetamol overdoses went up by more than others. In 1998 the Hong Kong media reported heavily on a case of carbon-monoxide poisoning by a very specific method, using a charcoal burner. In the ten months preceding the reports, there had been no such suicides. In the following month, November, there were three; then in December there were ten; and over the next year there were forty. You may want to remember that story for later.

It's not pie in the sky to suggest that the media should be careful about how they discuss suicide. After the introduction of media reporting guidelines in Austria, for example, there was a significant decrease in the number of people throwing themselves under trains.

So organisations like the Samaritans take this seriously. They suggest that journalists avoid crass phrases like 'a "successful" suicide attempt'. They suggest that journalists avoid explicit or technical details of suicide methods, for reasons you can now understand. They suggest that journalists include details of sources for help and advice, since an article about suicide represents a great opportunity to target people who are at risk with useful information. And they recommend avoiding simplistic explanations for suicide.

From the weekly mass of reports that trample on this perfectly good common sense, one article from the *Telegraph* at the tail end of last year particularly sticks in my memory. It is very different from the coverage of Sylvia Plath's son, and you might have missed it.

'Man Cut Off Own Head with Chainsaw' was the headline: 'A man cut off his head with a chainsaw because he did not want to leave his repossessed home.' What followed this headline was not a news story: far from it. What the *Telegraph* published was a horrific, comprehensive, explicit and detailed instruction manual.

In fact this information was so appallingly technical and instructive that after some discussion it has been decided that the *Guardian* will not print it, even in the context of a criticism. It gives staggering details on exactly what to buy, how to rig it up, how to use it, and even how to make things more comfortable while waiting for death to come. Suicidal thoughts are common. They pass.

Journalists get these kinds of stories from coroners' inquests, which are open to the public because we decided as a community, centuries ago, that it was important to be transparent about the judicial process.

Perhaps Sylvia Plath's son will have a public inquest. Perhaps the media will cover it in the same way that the *Telegraph* covered the tragic case of Mr Phyall. I doubt they will, and I very much hope they won't. It's just hard to tell which is the journalists' true voice: the caring, compassionate, informed consolation; or the murderously detailed chainsaw voyeurism.

Roger Coghill and the Aids Test

Guardian, 28 June 2008

It's the big stories I enjoy the most. 'Suicides "linked to phone masts"', roared the *Sunday Express* front-page headline this week. 'The spate of deaths among young people in Britain's suicide capital could be linked to radio waves from dozens of mobile phone transmitter masts near the victims' homes.'

Who is raising these concerns? 'Dr Roger Coghill, who sits on a government advisory committee on mobile radiation, has discovered that all 22 youngsters who have killed them-selves in Bridgend, South Wales, over the past 18 months lived far closer than average to a mast . . . Masts are placed on average 800 metres away from each home across the country. In Bridgend the victims lived on average only 356 metres away.'

These are extremely serious issues. Being generous, there is reasonable evidence of a possible link between power lines and childhood leukaemia, and we may not yet know the long-term physical risks posed by mobile phones to those who use them, since they haven't been around too long.

I contacted Dr Coghill, since his work is now a matter of great public concern: he will, quite naturally, want his evidence to be properly assessed. But Dr Coghill was unable to give me the data. No paper has been published. He himself would not describe the work as a 'study'. There are no statistics presented on it, and I am not allowed to see the raw figures. In fact, Dr Coghill tells me he has lost the figures. This is a bit off.

It also leads to obvious problems with interpretation: details are important, after all. He says the suicide rate was higher nearer mobile-phone masts: what was the control group he compared against? And how did he work out the average distance from a mast? Perhaps the average distance from a mast in any urban area is less than the average distance for the whole country, because masts tend to be clustered in urban areas, where the people are (like postboxes, or corner shops). Maybe densely populated poor areas with less political influence have more masts foisted upon them by planning committees, and maybe these poor areas also have more suicides.

Or maybe Dr Coghill is on to something? Clusters on maps have been the beginning of several interesting stories in epidemiology, including John Snow's discovering, in 1854, that the Broad Street pump was responsible for the Soho cholera outbreak. I asked Dr Coghill which 'averages' he meant. But he would not tell me.

Who is Dr Coghill? He says he doesn't have a doctorate, and that the *Express* made a mistake. Does he 'sit on a government advisory committee on mobile radiation'? Sort of. Mr Coghill participates in something called SAGE, a 'stakeholder' group which discusses power cables (not mobile phones) and is run at the request of the Department of Health by RK Partnerships Ltd, a company that specialises in mediation, facilitation and conflict resolution. People who campaign about things are rightly invited onto consultation panels run by the government, so that their concerns can be heard. I'm not sure if that makes them government advisers.

As an example of the kind of discussion you might find at SAGE, here is Mr Coghill's contribution to their last document, in the section where people who disagree with the group can state their own views: 'Whilst this first interim assessment is a welcome step, it contains three important omissions . . . the powerfully electro-protective effect of exogenous melatonin supplementation, particularly among the UK's 20 million elderly population, and the adverse effects of EMFs on melatonin synthesis within the body have not been addressed.' Mr Coghill recently received £125,000 of angel investment for his business selling a range of melatonin pills called Asphalia.

Readers worried by the front-page story on Mr Coghill's

inaccessible research may have visited his website for more information. There they could buy his electromagnetic field protection equipment at competitive prices, and a £149 device called the Acousticom for 'finding out if your home is being exposed to microwaves from e.g. cellphone masts', as well as several other interesting products, including a magnet that makes wine taste nicer, and the 'Mood Maker' treatment for impotence at just £22.32 including VAT ('The small unit discreetly attaches to your underwear . . . the Mood Maker will gently and gradually increase circulation in the pelvic area'). You might also enjoy his books, including *Electrohealing*, 'using electric and magnetic fields for alleviative and curative ends', and of course *Atlantis*, 'a new look at the Plato legend with a grim conclusion re global warming and ozone depletion'.

Lastly, regular readers will know that someone's ability to police their own enthusiasm can often be assessed using something called 'the Aids test'. Here is the *Express*'s front-page expert Mr Coghill on Aids: 'The idea that Aids is caused by a virus is a well-protected fiction.' Is there another cause? 'The possibility that immune deficits . . . can be acquired through over-exposure to non-ionising electromagnetic fields is, however, real, and proven in the laboratory.'

Because, remarkably, suicide is not the first problem Mr Coghill has attributed to electromagnetic waves, and he built his earlier hypothesis on the same evidence as his current one: 'Aids cases seemed to correspond closely to the numbers of RF, VHF, and UHF station densities.' Mr Coghill discovered that eleven of the twelve cities in America with the highest incidence of Aids also had the highest level of electromagnetic activity. A disease of dense urban areas, perhaps? He even had some exciting ideas about treatment: 'One first step might be to demagnetise the haem [sic] in an attempt to improve the signal to noise ratio of the immune signal . . .'

We should be glad that there are individuals out there with

esoteric views. We should respect and admire their tenacity and self-belief, if not their ability to provide us with actual data. But we probably don't need to put them on the front page of a national newspaper.

BRAINIAC

Ka-Boom! Science! COOL!!?!

Guardian, 22 July 2006

The new series of Sky TV's hit science programme *Brainiac* starts tomorrow, and there's just one question on everyone's lips: will they be faking the science as much in this series as they have previously?

At badscience.net you can find a very interesting clip from series two of *Brainiac*. It claims to show a lump of rubidium, and then a lump of caesium, each blowing up a bathtub: 'Whether you've left school, or you're still at school, you can still appreciate the sheer MAYHEM that chemistry can be!!!' the presenter explains. 'Bunsen burners! Mixing chemicals!' Science!!! 'Now, you may have been allowed to mix very small amounts of lithium with water.' Yes. 'You may, with a responsible adult, have mixed H_2O with sodium. And you may, under strict "scientific" control, have witnessed potassium mixed with water. But the odds are, if you have' – he reaches for a prop – 'it will only have been on those . . . rubbish science videos!' A box labelled 'Rubbish Science Video' is then burned on a Bunsen burner, using some big sciencey tongs.

Brainiac's much better than those boring science videos.

Then we get to the action. 'These next two are the dog's nuts of the periodic table,' he goes on, introducing rubidium and caesium. 'Mix these two with water, stand back . . . and watch the MAYHEM!!!' In the programme they are explicit about what they are doing. 'Caesium, the emperor of alkali metals, particularly nasty, could go off at any time!' 'What's that going to do when it hits the water?' 'Imagine a depth charge in a bathtub!'

That's exactly what I'm looking at. I can see the black wire, connected to the detonator.

'As our caesium sinks in the water, the rapid generation of hydrogen gas should produce quite an explosion!!!' They drop the caesium in, and run for cover. 'And it does!' The bathtub is blown to pieces. 'Yeeeeeesssss!' gasps the presenter. 'Only on *Brainiac* do you get that kind of . . . science!' The dance music stops.

But what really happened? I have a Deep Throat: *Brainiac*'s 'Dr Bunhead', also known as Tom Pringle. He says: 'Absolutely bloody nothing. The density of caesium ensured it hit the bottom of the bath like a lead weight. The sheer volume of water then totally drowned out the thermal shock-wave I was expecting to shatter the bath. This was an expensive filming day. They had hired part of Pinewood studios and had an ambulance and fire engine plus crew on standby. They could not go home empty-handed. So they rigged a bomb in the bottom of the bath (you really can see the black wire leading into the bath, if you watch the show again) and then blew the shit out of it. I must say, it did look cool, [but it] ate away at my conscience . . . I couldn't do anything about it.'

If this was all faked, then what's the point? Or rather, why do they make such a fuss about how they're really doing science, and saying 'the rapid generation of hydrogen gas' caused the explosion, if it didn't? Anyway, a Sky spokesperson said: 'All of the experiments conducted on *Brainiac* have proven theory behind them. We aim to inform, excite and, above all, entertain our viewers with science method conducted in a fun and engaging way. We love big bangs and sometimes we'll make an explosion bigger than we need to just because it's fun but we always tell our viewers. We're just about to start our fourth series, we've won several awards as well as the respect of educational professionals and we're really proud to be sparking children's interest in science.' Sky say they tell people when experiments are souped up. They were unable to tell me if this experiment was faked or not (it was), and unable to confirm if viewers were told if it was faked (they weren't).

Who's the Daddy?

Guardian, 22 July 2006

So Theodore Gray bought a kilo and a half of pure sodium metal on eBay. At school, you probably dropped a crumb of it into water – or rather, you watched your chemistry teacher do that – and the sodium reacted with the water to produce sodium hydroxide (a nasty alkali) and some hydrogen gas. The reaction gave off lots of heat, which ignited the hydrogen, and so the little lump of sodium fizzed across the water with a nice flame.

Theodore Gray got some friends over, with refreshment, and launched a kilo of sodium into his private lake.

His reasoning was sound: if he tipped in some hydrochloric acid afterwards ('Muriatic acid at any hardware store'), this would neutralise the sodium hydroxide, and the pond would be a little saltier. There's no law against making slightly salty water.

That's not quite how things worked out. After an initial large explosion from the first chunk, a series of secondary explosions occurred, producing one fairly large wedge that began hopping across the lake. It was thrown forty feet up into the air, then flew into the water at high speed, only to be thrown back into the air by the resulting explosion. It only takes a few of these skips to get several hundred feet in a few seconds. The party-goers were two hundred feet away, and ran for cover.

Now, you might be asking: where's the bad science? Well: Sky's popular flagship science programme has just started its new series. Last week I accused them of faking content. They tried to make me nervous about it. Now they've admitted that they definitely did fake those explosions. And they have also admitted that viewers were not told (or as they said last week: 'but we always tell our viewers'). And they have admitted that they fake other stuff. In fact, they were so blasé about this that at one point they were even going to give me a list of other examples, but now they've changed their mind about that.

Here's where it gets really elaborate: they don't tell you explicitly that they fake stuff, but they now say that you are a fool not to assume that they fake their experiments: 'The clue is in the title of the show, "*Brainiac Science Abuse*", it's an entertainment programme, it's being made for an entertain-ment channel, it's to be expected from the show.'

But it's not. This is a programme that repeatedly tells viewers how reckless and dangerous and science it is – in a way that now feels slightly defensive. Now *Brainiac* claim they actually said 'This is what happens if you stick rubidium in a bath,' and then showed 'a demonstration of what would happen' (that's just not true: it was a generic special effects explosion, and they said – repeatedly – that they were doing it for real). 'We may as well have done it,' they say: which is an interesting approach to science. But of course, they *did* do it: their scientific adviser dropped these metals into their bath, on camera, and unfortu-nately the bath didn't blow up. That's life. You can't say that this 'not exploding' was somehow 'wrong', and that the fake explosions they broadcast were what 'should' have happened. What should have happened, when you drop the rubidium in the water, is exactly what did happen: not a lot.

Despite the fakery, of course, *Brainiac* gets massive ratings, and is praised in very high places for popularising science. So to me, this is a lot like the nutritionist question: Is it OK to lie to people about science, if it makes them eat vegetables? But more than that, it's a question of who do you want to be your friend: the faker, who desperately insists he's doing dangerous science, while setting off weak, staged, plastic explosions; or Theodore Gray, who buys a kilo and a half of sodium on the internet, and gets some friends over for a party, to chuck it in the lake?*

---

* After this column, Theodore Gray bought some five-gram chunks of caesium and rubidium – much more than *Brainiac* didn't use – and threw them into water in his garden. In the video, you can see some light, some pinging, and some phutting, but no explosion. Why not? Although caesium and rubidium are technically more reactive than sodium, he explains, in reality you get a bigger bang for your money from sodium: the atoms are smaller, so you get more atoms per gram, so the same-sized lump makes more hydrogen. 'Under typical night-time escapade conditions, the larger hydrogen explosion created by sodium more than makes up for the more vigorous initial decomposition reaction of caesium. It's a pity that *Brainiac* felt they needed to perpetuate a myth by faking it, when the truth is even better: common everyday sodium beats out those high-priced exotic elements.'

STUFF

Here's My . . . Foreword to the Romney, Hythe and Dymchurch Railway Guidebook

20 December 2013

I often tweet about my love for the RH&D narrow gauge railway, and this year they asked me to write an introduction for their guidebook. Some of the staff were worried by what I sent. But they were wrong. I love this railway.

The Romney, Hythe and Dymchurch Railway has a strange, dreamlike existence, on the border between fantasy and reality. You leave Toytown in a cute miniature train, surrounded by excited children. But Disney this is not. Suddenly you're riding through real life: past clothes lines, collapsing breezeblock walls, an abandoned washing machine in a back garden, chuffing along behind a miniature steam train. Finally, you're ferried across a beautiful, windswept shingle peninsula, spotted with railway-carriage houses and abandoned shipping containers. Then you are delivered to the foot of a nuclear power station.

When I mentioned that huge, monolithic nuclear installation to the editor of this guidebook, he replied: 'Do you know, I barely notice it these days.' The two reactors are at least a hundred metres tall, humbling and majestic, on a 225-acre site. Nobody should ever play them down, and it's fun to explore the perimeter.

This meeting of toy train sets and grim industrial purpose is what makes the Romney, Hythe and Dymchurch Railway so perfect. Toytown trains in amusement parks are annoying. Proper narrow-gauge railways have a history. When you stop being a child, and start herding children yourself, you notice the chinks in playtime. Now, when I see kids chasing each other through a carriage – on a rebuilt Welsh mining railway, say – I think: this dragged people to hard jobs, in dark pits, that most people today could barely imagine.

For the Romney, Hythe and Dymchurch Railway, the battle between childish fantasies and real life is even denser. It was conceived by two wealthy men, each obsessed with miniature railways, because they wanted their own to play with. Both were racing drivers, and one was a real-life Count. In the war it was used for serious freight work, while the world's only miniature armoured train trundled up and down the track, monitoring the coast for invasion by the Germans. For the soldiers stationed here, amidst all the terror and uncertainty of war, it must have seemed like a very peculiar dream.

I've often fantasised about the same thing myself – combining grim reality and a very big train set – by moving to Dungeness. I could handle the travel time, five days a week, if the commute made adult life seem like a huge, silly game: through the fog, past the nuclear power station, waiting at a miniature station for my tiny train, then up to the mainline and on to meetings. All the while I'd know that my home, come nightfall, was a converted railway carriage, on a pile of shingle, by a nuclear power station, at the end of Toytown. Welcome to this very peculiar, working paradise.

If you like nerdy day trips, you should visit www.nerdydaytrips.com, a crowd-sourced website I built with friends, filled with thousands of weird days out from all over the world (although it's recently been infested with normal days out too, after someone posted it on a museum curators' mailing list). It's great for planning a comprehensive nerd itinerary into any journey. For example, when you visit Dungeness for the RH&D Railway, it is compulsory to grow a WWII moustache and stop in at the sound mirrors, just one mile away. As you can see, when I did this, I found it all extremely exciting.

How I Stalked My Girlfriend*

Guardian, 1 February 2006

For the past week I've been tracking my girlfriend through her mobile phone. I can see exactly where she is, at any time of day or night, within 150 yards, as long as her phone is on. It's been very interesting to find out about her day. Now I'm going to tell you how I did it.

But first: my girlfriend is a journalist, I had her permission ('In principle . . .'), and this was all in the name of science. You have nothing to worry about, at least not from me.

First I had to get hold of her phone. This wasn't difficult. We live together and she has no reason not to trust me, so she often leaves it lying around. I needed it for five minutes, to register it on a website I'd been told about. It looks as if this service is mainly for tracking stock and staff movements: I've shown the *Guardian* staff that it works, but they won't let me tell you the name of the site (I agree). I ticked the website's terms and conditions without reading them, put in my debit card details, and bought twenty-five GSM credits for £5 plus VAT.

My girlfriend's phone vibrated with a new text message: 'Ben Goldacre has requested to add you to their Buddy List! To accept, simply reply to this message with "LOCATE".' I sent the reply. The phone vibrated again. A second text arrived: 'WARNING: [This service] allows other people to know where you are. For your own safety make sure that you know who is locating you.' I deleted both these text messages, and put the phone back in my girlfriend's bag.

On the website, I see the familiar number in my list of 'GSM

---

* I still get a handful of emails every year from creepy men asking me how to do this.

devices', and click 'Locate'. A map appears of the area where we live, with a person-shaped blob in the middle, roughly a hundred yards from our home. The phone doesn't make a sound. It gives no sign of what I'm doing.

I can't quite believe my eyes. I knew the police could do this, and phone companies, but not any creepy boyfriend with five minutes' access. There is nothing on my partner's phone that could possibly let her know that I'm tracking her location. I set the website to record her location at regular intervals, and plot her path on the map, so I can view it at my leisure. Even with her permission, it felt very wrong.

By the time she got home, I was over-excited, and the secret lasted less than a minute. To my disappointment, she wasn't freaked out. But then, I already knew that she hadn't gone to her ex's flat, and she'd only really been at work all day, apart from one trip to the bank. It felt strangely protective, looking down on her, following her silently. If there's anyone who might want to track you, and they've had access to your phone, call your phone company, create havoc, and make them find out if there's a trace on your phone.

EARLY SCARES

EARLY SNARKS

Staying Beautiful Is Easy to Do

Guardian, 1 May 2003

It's been a great week for Bad Science spotters. Bob Conklin writes in about Seasilver nutrient potion. Kirlian photography of your aura will demonstrate an 'increase in energy' after taking it, and just one capful will deliver 'EVERY [sic] vitamin, macro mineral, trace mineral, amino acid, enzyme, and bio-element known to man' straight to your system. As Bob says: 'I'm not sure I want every enzyme and bio-element known to man in my mouth.'

Dr Victoria Kaziewicz sends us even more preposterous pseudo-science: 'In a book called *How to be Beautiful*, by Kathleen Baird Murray, you can read that beauty products containing natural ingredients are preferable because naturally occurring substances are irregularly shaped like the substances making up your own body, while manufactured chemicals are perfect spheres.'

Dr Cicely Marston writes about an easier way to stay beautiful. A team from Harvard School of Public Health took six years to find that watching television for an extra two hours a day increased the rate of obesity by 25 per cent in 50,000 women. Magnificently obvious – but possibly less obvious is why they were only looking at women.

Picky Bad Science Spotter of the Week Award goes to Jennifer Leech, who has been bothered for decades by an issue in *Lord of the Flies*. 'In the book it says that Piggy has myopia. So,' she continues, 'how can the children marooned on that island have used his glasses to start a fire?'

There's hope on the horizon for the so-called Sars epidemic (as opposed to malaria, which kills a million people a year, and tuberculosis, which kills three million). Richard Spacek sent us a full-page ad from Canada's *Saturday National Post* from the Dr Rath Health Foundation: 'It is a scientific fact that all viruses that have been scientifically investigated can be blocked by specific natural essential nutrients.' The fact that life-saving information 'is being withheld from the people of the world is irresponsible and must be stopped immediately'. Never let it be said that I am part of any global conspiracy to suppress this vital information.

And finally, thank God that in this cynical world, in Wales, Dewi ap Ifan is still managing to feel optimistic: 'Aren't we all lucky that Sars has arisen in China? Traditional Chinese medicine, herbalism and acupuncture will have it under holistic control in no time.'

Keep the Bad Science coming: you are not alone.

Because You're Worth It

Guardian, 27 November 2003

Reader Helen Porter writes in to tell me about the Ion-Conditioning Hairdryer, which uses 'Patented Trionic Action' to 'micronize' water molecules and, impressively for a hairdryer, magically hydrate your hair. The *Journal of Trionic Physics* – in case you thought the hairdryer people made those words up themselves – was actually the name of a Jefferson Airplane fanzine. But I digress: the manufacturer, Bio-Ionic, is also the inventor of Ionic Hair Retexturising (IHR). And this is not just a new way to straighten your hair, it's a whole new branch of physics.

Colour Nation, hairdressers to the stars in Soho, London, offers Bio-Ionic's IHR. Its public-relations material explains how it works: 'Positive ions have lost an electron, and are considered unhealthy,' whereas negative ions 'have gained an electron, and greatly assist in a body's mood, energy level, and overall health'. When these benevolent negative ions encounter water, 'the water molecules are broken down to a fraction of their previous size . . . diminutive enough to penetrate through the cuticle, and eventually into the core of each hair'.

I might be wrong, but surely shrinking water molecules must cost more than the £230 Colour Nation charges for IHR? The only other groups that have managed to create that kind of super-dense quark-gluon plasma used a relativistic heavy ion collider, and if Colour Nation has got one of those at the back of the salon, then I'm glad I don't live in the flat upstairs. Although a *Mirror* reporter who had the compressed molecule treatment did say her hair 'itched and smelled of chemicals' afterwards. Maybe there's something more potent than negative ions in there after all.

Meanwhile, a tip from a friend – who, may I just point out, doused for the sex of her baby. She was delighted, at her ante-natal yoga class, after being told how immunisations would kill her baby, to be handed *Homeopathy News*. The pamphlet mentions a study from the Royal London Homeopathic Hospital in which 80 per cent of twenty-five children reported an improvement in their asthma after homeopathy. Which sounds impressive. But there was no placebo control group, and it doesn't seem to have actually been published anywhere (or not anywhere peer-reviewed). Which doesn't mean it's not true. Just remember that in a recent review of all the evidence on homeopathy – I'll say it again – it was shown, overall, to be no more effective than a placebo . . .

More Than Water?

Guardian, 22 January 2004

What is it with pseudo scientists and water? After last week's 'clustered water', Caroline Stacey in the *Independent*'s Food and Drink section was getting excited about Oxygizer water. 'Oxygizer doesn't just slake a thirst, it provides the body with extra oxygen too. A litre contains 150mg of oxygen, around twenty-five times more than what's in a litre of tap water.' Handy. 'This apparently helps remove toxins and ensures a stronger immune system, as well as assisting the respiratory system, so you recover better from exercise . . . cleverly they've added something to water that's not an additive.'

In the spirit of Victorian gentleman scientist self-experimenters, I decided to put Oxygizer to the test. Back in the sixties, a scientist in New York managed to get mice breathing underwater, in a saline solution at six times normal atmospheric pressure, just like in that movie *The Abyss* – it takes a lifetime of popular-science books to collect this kind of trivia. Unfortunately, the mice died after eighteen hours.

So instead, I decided to drink some Oxygizer after a three-mile run. I can take in about 100ml of oxygen with every breath, or 150mg, and like most humans I only absorb about 30mg of that. That's 300mg a minute, but after serious exercise it goes up to about 3,000mg a minute. To help myself recover significantly faster after my run, I figured I'd need an extra 20 per cent of oxygen, or 600mg a minute. That meant drinking forty litres of Oxygizer over ten minutes, getting the stuff into me at a rate of one litre every fifteen seconds. This cost £120, and almost doubled my body weight, but it's all in the name of science. Fairly soon my circulatory system was so overloaded that pints of frothy sputum were spewing up at the back of my throat; then my abdomen burst open and covered my laptop in guts.

This was inconvenient. But with my dying keystrokes, just in case any Oxygizer customer ever looks up Oxygizer to buy some Oxygizer, I'm using the word Oxygizer as much as Oxygizer, because articles on the *Guardian* website come up right at the top of the first page on Google keyword search.

'Nanniebots' to Catch Paedophiles

Guardian, 25 March 2004

As I sit here, quietly shedding the weight off my fat arse in my Dr Norbert Wurgler caffeine-impregnated SlimFit tights, I find myself bitterly regretting the title of the column. Here's one I'm not sure of. Artificial intelligence, in the form of 'Nanniebots', is being used to catch paedophiles. Nanniebots are AI programs which hang out in internet chatrooms, allegedly spotting the signs of grooming. They have done 'such a good job of passing themselves off as young people that they have proved indistinguishable from them', according to *New Scientist*. So that's the Turing test – where a computer program is indistinguishable from a real person – passed; and who'd have thought it, in a program written by a lone IT consultant from Wolverhampton with no AI background. So I call him.

Here's the problem. Reading *New Scientist*'s chat with Nanniebot, the excellent www.ntk.net/ (*Private Eye* for geeks) points out that Nanniebot 'seems to be able to make logical deductions, parse colloquial English, correctly choose the correct moment to scan a database of UK national holidays, comment on the relative qualities of the Robocop series, and divine the nature of pancakes and pancake day'. Jabberwock, the winner of last year's Loebner Prize for the Turing test, is rubbish in comparison (you can talk to it online and see for yourself). But Jim Wightman, the Nanniebot inventor – whose site claims they've passed the Turing test – isn't entering the Loebner Prize this year. Maybe next year . . . it's too buggy. But it's live on the internet already? Can I test it? Sure. But I want to see with my own eyes that there's not a real human being somewhere tapping out the answers, I explain. Jim offers network-monitoring software on my computer, to prove it's connected to the one server. But what about that server? I want to see it working on its own, without a human. Can I come round to Jim's place? He chuckles . . . Jim doesn't keep the conversation datasets on site in Wolverhampton. 'I know it sounds a bit *Mission Impossible*, but . . .' He's worried they might get stolen. They're in a secure facility 'with an iron lid under a mountain'. He has no copies. It's eighteen terabytes of data, he says. There are copies in the hosting facilities, and one in London. I offer to go there. 'There might be security issues with them letting us in,' he says. So here it is. I'm going if I can. I'd love to see it work. If there is an AI academic who wants to come, email me: it could be the biggest ever break-through in AI. Or it could be a lot of fun.

Nanniebots and Neverland

Guardian, 1 April 2004

Where were we? Everyone was questioning the authenticity of Jim Wightman's paedophile-entrapping artificial-intelligence chat program Nanniebot, since it was more than ten years ahead of all other artificial-intelligence technology, and no one is allowed to see it in the flesh. But Jim – from the unfortunately named Neverland Systems – had personally guaranteed me a demonstration. Now he has changed his mind, although he is still claiming to have thousands of Nanniebots in action on the web. I'm not going to waste your time with any analysis of his 'chat transcripts', since no one can be sure they were generated by his program.

Of course, the BBC, ITV and *New Scientist* couldn't possibly have known that Jim was caught out making false claims about writing software a year ago, on the Holocaust denial newsgroups he likes to frequent. He now admits to making these false claims, but said they were made in jest. He also got noticed in the Tivo hacking discussion boards, claiming to have modified the device to stream shows over a network, which the other experts felt was impossible. Jim provided no evidence to make them think otherwise, and disappeared. He still claims to have it working.

People are perfectly entitled to spend time on Holocaust denial chatboards if they really want to. Jim admits posting as 'Death's Head', the same name as the SS murder and torture units. Death's Head has made postings containing violent and graphic threats to rape, assault and kill, often with a firearm, in the context of chatboard discussions about the Holocaust.

In an online discussion after similar violent threats were mentioned, a posting did state that 'me = Jim Wightman = Death's Head = Totenkopf . . . all you needed to do was ask'. Jim denied to me that he made the postings, and says they were faked. Maybe they were, but his previous postings give reason to question his work. So far, he's made a grand claim with no good evidence: business as usual for Bad Science.

Now he's collecting donations and volunteers for chatnannies.com, a service where adults will enter children's chatrooms to monitor them for paedophile activity. He will now be greatly assisted in this venture by the fact that he can cite on his website the uncritical reports on his work by *New Scientist* and the BBC.

Artificial Intransigence

Guardian, 24 June 2004

You may remember Jim Wightman. He claimed to have written a piece of chat software that could pass itself off as a real child in a chatroom, and identify internet paedophiles by their behaviour. To say this was thought highly dubious is an understatement: the software, if it existed, would have been ten years ahead of everything written by huge teams of AI academics; he offered to let us see the software working, and then refused; and the NSPCC and Barnardo's distanced themselves from his ideas about monitoring children's activities, partly because he has no child-protection background. Embarrassingly, *New Scientist* accepted his claims uncritically, and the BBC and others followed suit, although *New Scientist* did, after two pieces here, remove its glowing article about him from its website.

Now they're back with Wightman. Here's what happened. *New Scientist* visited Jim at home with two AI academics to chat with the program. In previous 'test conversations' over the web – where experimenters couldn't tell if the computer was working alone – the program gave highly sophisticated answers, albeit after a rather long delay (almost as if someone was typing them). This time, running on a machine in the room, with Jim present, it instantly gave rubbish computer-generated responses, that were nothing like those in the previous transcripts. In fact, it gave the very same answers that 'Alice'

– an old and not very sophisticated AI program written by somebody else, not Wightman – gave in subsequent tests. Then Wightman offered to show them the code. But suddenly, and inexplicably, the power to Jim's whole house went off. The test was over.

Did *New Scientist* finally give it up? No. '*New Scientist* can still provide no definitive proof of Wightman's claims, but looks forward to a return visit when the complete ChatNannies software is available for testing.' Please. Did it ask Wightman about his claim to have a seven-figure offer from an American corporation which had 'full independent testing performed on the AI and are confident of its validity and effecacy [sic]'? He was, apparently, quite capable of giving *them* a proper demonstration. Did it quiz Wightman on his previous false claims about writing software, or any of the other issues Bad Science raised? No. To those of us brought up loving the great institution of *New Scientist* this is – as Tibor Fischer said in that famous book review – a bit like bouncing out of the classroom at breaktime, only to catch your favourite uncle masturbating in the school playground.

BOOKENDS

Be Very Afraid: The Bad Science Manifesto

Guardian, 3 April 2003

It was the MMR story that finally made me crack. My friends had always seemed perfectly rational: now, suddenly, they were swallowing media hysteria hook, line and sinker. All sensible scientific evidence was twisted to promote fear and panic. I tried to reason with them, but they turned upon me: I was another scientist trying to kill their baby.

Many of these people were hardline extremists, humanities graduates, who treated my reasoned arguments about evidence as if I was some religious zealot, a purveyor of scientism, a fool to be pitied. The time had clearly come to mount a massive counter-attack.

Science, you see, is the optimum belief system: because we have the error bar, the greatest invention of mankind, a pictorial representation of the glorious, undogmatic uncertainty in our results, which science is happy to confront and work with. Show me a politician's speech, or a religious text, or a news article, with an error bar next to it.

And so I give you my taxonomy of bad science, the things that make me the maddest. First, of course, we shall take on duff reporting: ill-informed, credulous journalists, taking their favourite loonies far too seriously, or misrepresenting good science, for the sake of a headline. They are the first against the wall.

Next we'll move on to the quacks: the creationists, the new-age healers, the fad diets. They're sad and they're lonely. I know that. But still they must learn. Advertisers, with their wily ways, and their preposterous diagrams of molecules in little white coats: I'll pull the trigger.

And the same goes for the quantum spin on government science. I'm watching you all.

And finally, let us not forget the strays, the good scientists who have passed to the dark side. Was it those shares in that drug company, or the lust for fame and glory? Bad scientists, your days are numbered.

If you are a purveyor of bad science, be afraid. If you are on the side of light and good, be vigilant; and for the love of Karl Popper, email me every last instance you find of this evil. Only by working joyously together can we free this beautiful, complex world from such a vile scourge.

What Eight Years of Writing the Bad Science Column Has Taught Me

Guardian, 4 November 2011

I've got to go and finish a book: I'll be back in six months, but in case it kills me, here's what I've learned in eight years of writing this column.

Alternative therapists don't kill many people, but they do make a great teaching tool for the basics of evidence-based medicine, because their efforts to distort science are so extreme. When they pervert the activities of people who should know better – medicines regulators, or universities – it throws sharp relief onto the role of science and evidence in culture. Characters from this community who wonder why people keep writing about them should look at their libel cases and their awesomely bad behaviour under fire. You are a comedy factory. Don't go changing.

Next: the real story of how the world works is much weirder than anything a quack can make up. The placebo effect is maddening, the nocebo effect more so, but the research on how we make decisions, and are misled by heuristics and mental shortcuts, is the wildest of all. Knowing about these belief-hacks gives you thrills, and power.

Pharmaceutical companies can behave dismally. Most important, they still won't publish all the results of all the clinical trials conducted on humans. This is indefensible, and because we tolerate it, we don't know the true effect sizes of the medicines that we give. This absurd situation mocks the whole of medicine: we need legislation to fix it, and popular movements to drive that. I'll join yours.

Journalists can mislead the public about the answers of evidence-based medicine, which is bad. But they also mislead us on the methods and techniques. We live in a new era of doctors and patients – at our best – making decisions together. For that collaboration to work, everyone needs to understand how we know if something is good for us, or bad for us. The basics of evidence-based medicine, of trials, meta-analyses, cohort studies and the like should be taught in schools and waiting rooms. It's interesting, but it's also life and death: people care about it.

Politicians misuse evidence, and distort it to shameful degrees. But more than that, there are endless cases where we could do randomised trials on policies – old and new – to find out if they achieve the outcomes they're aiming for. There is no honourable excuse for failing to use the fairest tests we can design.

Real scientists can behave as badly as anyone else. Science isn't about authority, or white coats, it's about following a method. That method is built on core principles: precision and transparency; being clear about your methods; being honest about your results; and drawing a clear line between the results, on the one hand, and your judgement calls about how those results support a hypothesis. Anyone blurring these lines is iffy.

Conflict-of-interest stories – where someone has a vested interest in the results of their study – are important, because they tell you when there's a risk that something's wrong in a piece of science. But this is only motive: the gruesome, fascinating mechanism of a crime against science – the methodological flaws – that's where the action is. People who don't really understand science can only critique it in terms of motive. Let them have that; we'll do the details.

Last, nerds are more powerful than we know. Changing mainstream media will be hard, but you can help create parallel options. More academics should blog, post videos, post audio, post lectures, offer articles, and more. You'll enjoy it: I've had threats and blackmail, abuse, smears and formal complaints with forged documentation.

But it's worth it, for one simple reason: pulling bad science apart is the best teaching gimmick I know for explaining how good science works. I'm not a policeman, and I've never set out to produce a long list of what's right and what's wrong. For me, things have to be interestingly wrong, and the methods are all that matter.

So keep the nonsense coming, I'll see you next year for more, and if you miss me, I'll be procrastinating at badscience.net, and @bengoldacre on Twitter.

I haven't yet gone back to a weekly column. The things I write take a huge amount of time: not just for the research, but also from chasing down endless blind alleys to find that one shining gem of bad behaviour that can illustrate an interesting bit of science.

To reassure you, I haven't been smoking dope on the sofa. Over the past three years I've written Bad Pharma, *and pushed hard on the policy fallout that it triggered (which you can read about in the updated second edition). I've fallen back in love with seeing patients; dived headlong into teaching; had exciting, busy day jobs; done hundreds of talks; put fingers into pies; and fallen into lobbying and policy work on the changes I've advocated for (which you can read about in the last book and, especially, the next one). Lastly, I've become obsessed with the technical fun of shaping a big argument over a book rather than 700 words: there are two of these to come very soon. Also, we've had babies.*

I am extremely optimistic about the growing role of science in society. There is still an endless stream of nonsense in mainstream media, politics, and medicine: but the last ten years have seen a spectacular flourishing of science, in live shows, and most importantly online. Rebuttals, fact-checks and evidence can be thrown onto Twitter or blogs within minutes of a dubious claim being made, and anyone can get access to good quality information, wherever they're motivated.

If you are a purveyor of bad science: the mountain of bullshit is slowly shifting beneath your feet. You should be worried.

And if you are a nerd: keep talking, stay vigilant, and be proud. This is our time.

Acknowledgements

I have been lucky enough to be taught, corrected, calibrated, cajoled, amused, housed, helped, loved, reared, encouraged and informed by a very large number of smart and excellent people, including (each, to be clear, for only a subset of the preceeding activities): Liz Parratt, John King, Steve Rolles, Mark Pilkington, Shalinee Singh, Emily Wilson, Ian Katz, Iain Chalmers, Alex Lomas, Liam Smeeth, Ian Sample, Carl Heneghan, Richard Lehman, Kathy Flower, Ginge Tulloch, Matt Tait, Carl Reynolds, Dara Ó Briain, Paul Glasziou, Simon Wessely, Cicely Marston, Archie Cochrane, William Lee, Hind Khalifeh, Martin McKee, Cory Doctorow, Evan Harris, Muir Gray, Rob Manuel, Tobias Sargent, Anna Powell-Smith, Tjeerd van Staa, Robin Ince, Fiona Godlee, Trish Groves, Tracy Brown, Sile Lane, David Spiegelhalter, Ute-Marie Paul, Roddy Mansfield, Amanda Palmer, Rami Tzabar, George Davey-Smith, Charlotte Wattebot-O'Brien, Patrick Matthews, Amber Marks, Giles Wakely, Andy Lewis, Suzie Whitwell, Harry Metcalfe, Gimpy, David Colquhoun, Louise Burton, Simon Singh, Vaughan Bell, Nick Mailer, Milly Marston, Tom Steinberg, Mike Jay, Chris, Tom, Reg, Mum, Dad, Josh, Raph, Allie, Archie, Alice and Lou. I'm hugely indebted to the late Pat Kavanagh, Zoe Ross, Rosemary Scoular and especially Sarah Ballard. Robert Lacey and Louise Haines at 4th Estate are mighty and strong.

My entire brain is now outsourced to a synchronised information monster fashioned from Evernote, Zotero, InstaPaper, Feedly and Twitter, and all plumbed in together using the high-level programming service If This Then That. This makes me more happy, less bored and more productive than I could possibly imagine being in any other era of human history. Scrivener will change your life if you write long structured documents. IntervalTimer gives you twenty-minute bursts of work followed by five of dithering. AntiSocial is a piece of software that, on a timer, irreversibly disables Twitter and Gmail on your computer when you're working. Use this information wisely. In recent years I've had day jobs in various places including the magnificent London School of Hygiene and Tropical Medicine and endless hospitals, supported by the National Institute for Health Research, the Scott Trust, the Wellcome Trust, Nuffield College Oxford, and the NHS, and a bursary from the Oxford University Business Economics Programme.

Notes

HOW SCIENCE WORKS

Why Won't Professor Susan Greenfield Publish This Theory in a Scientific Journal?

Why Won't Professor: http://www.badscience.net/2011/11/why-wont- professor-greenfield-publish-this-theory-in-a-scientific-journal/

announced that computer games: http://www.thesun.co.uk/sol/homepage/woman/health/health/3871474/Computer-games-are-giving-kids-dementia.html

dementia in children: http://www.dailymail.co.uk/health/article-2049040/Computer-games-leave-children-dementia-warns-neurologist.html

not really what she meant: http://beefjack.com/news/the-sun-misrepresented-scientist-in-games-dementia-article/

rise in autism diagnoses: http://www.newscientist.com/article/mg21128236.400-susan-greenfield-living-online-is-changing-our-brains.html

then pulled back: http://www.dailymail.co.uk/sciencetech/article-2023535/Battle-dons-internet-link-autism-scientists-claim-PCs-shorten-attention-span.html

autism charities: http://www.guardian.co.uk/society/2011/aug/06/research-autism-internet-susan-greenfield?CMP=twt_fd

Oxford professor of psychology: http://deevybee.blogspot.com/2011/08/open-letter-to-baroness-susan.html

They seem changeable: http://www.guardian.co.uk/science/the-lay-
scientist/2011/aug/08/1

derided in the media as sexist: http://www.thisislondon.co.uk/standard/
article-23793960-the-male-rage-that-is-a-bad-rap-for-science.do

Professor Greenfield responded: http://www.guardian.co.uk/society/
2011/aug/06/research-autism-internet-susan-
greenfield?CMP=twt_fd

Cherry-Picking Is Bad. At Least Warn Us When You Do It

Cherry-Picking: http://www.badscience.net/2011/09/cherry-picking-
is-bad-at-least-warn-us-when-you-do-it/

Aric Sigman: http://www.guardian.co.uk/commentisfree/2011/
sep/14/daycare-cortisol-levels-children?INTCMP=SRCH

Professor Dorothy Bishop: http://deevybee.blogspot.com/2011/09/
how-to-become-celebrity-scientific.html

Sigman himself admits it: http://www.badscience.net/2011/09/2009/
02/the-evidence-aric-sigman-ignored/

pdf on his website: http://www.aricsigman.com/IMAGES/Statement.
pdf

Celia Mulrow: http://www.bmj.com/content/309/6954/597.full

'systematic reviews': http://www.bmj.com/content/315/7109/672.full

the last time: http://www.badscience.net/2011/09/2009/02/the-
evidence-aric-sigman-ignored/

deliberately incomplete article: http://www.youtube.com/watch?v=
Gg8LlUME-IM&feature=player_embedded

Being Wrong

Being Wrong: http://www.guardian.co.uk/theguardian

research about research: http://archinte.ama-assn.org/cgi/content/
short/archinternmed.2011.295

John Ioannidis: http://jama.ama-assn.org/cgi/pmidlookup?view=lon
g&pmid=16014596

Kids Who Spot Bullshit, and the Adults Who Get Upset About It

Kids Who Spot: http://www.badscience.net/2011/06/kids-who-spot-
bullshit-and-the-adults-who-get-upset-about-it/

Ryan Giggs's penis: http://en.wikipedia.org/wiki/Ryan_Giggs
Brain Gym: http://www.badscience.net/2011/06/category/brain-gym/
writing about since 2003: http://www.badscience.net/2011/06/2003/06/
    work-out-your-mind/
pay hundreds of thousands: http://www.davidcolarusso.com/blog/?
    p=48
hundreds of state schools: http://www.google.co.uk/search?q=%22br
    ain+gym%22+inurl%3Asch.uk
Emily Rosa: http://en.wikipedia.org/wiki/Emily_Rosa
published a scientific paper: http://jama.ama-assn.org/content/279/13/
    1005.full
Journal of the American Medical Association: http://en.wikipedia.org/
    wiki/Journal_of_the_American_Medical_Association
practitioners were deeply unhappy: http://en.wikipedia.org/wiki/
    Emily_Rosa
Rhys Morgan: http://www.guardian.co.uk/science/2010/sep/15/
    miracle-mineral-solutions-mms-bleach
www.crohnsforum.com: http://www.crohnsforum.com/
finding official documents: http://www.fda.gov/Safety/MedWatch/
    SafetyInformation/SafetyAlertsforHumanMedicalProducts/
    ucm220756.htm
The adults banned him: http://thewelshboyo.co.uk/?p=87
The One Show: http://www.youtube.com/watch?v=R0yUYOz62wk
Chief Medical Officer for Wales: http://thewelshboyo.co.uk/?p=54
have waded in: http://thewelshboyo.wordpress.com/2010/10/01/
    bleachgate-more-on-mms-in-the-uk/
evidence-based medicine: http://www.jameslindlibrary.org/testing-
    treatments.html
tell the world: http://wordpress.com/

Existential Angst About the Bigger Picture
Existential Angst: http://www.badscience.net/2011/05/existential-
    angst-about-the-bigger-picture/
major review in 2003: http://faculty.virginia.edu/haidtlab/jost.glaser.
    political-conservatism-as-motivated-social-cog.pdf
study from 2004: http://psp.sagepub.com/content/30/9/1136.
    abstract

new study: http://www.plosone.org/article/info%3Adoi%2F10.1371%2
Fjournal.pone.0017349

most recent estimate: http://www.stratongina.net/files/50million
ArifJinhaFinal.pdf

systems are imperfect: http://www.badscience.net/2011/05/category/
regulating-research/

The Glorious Mess of Real Scientific Results

Glorious Mess: http://www.badscience.net/2010/11/the-glorious-mess-
of-real-scientific-results/

Solomon Asch's legendary studies: http://en.wikipedia.org/wiki/
Asch_conformity_experiments

hints of patterns: http://psycnet.apa.org/?&fa=main.doiLanding
&doi=10.1037/0033-2909.119.1.111

a new variant: http://www.informaworld.com/smpp/section?content
=a922516755&fulltext=713240928

sometimes happens with uncomfortable data: http://www.badscience.
net/2009/08/how-myths-are-made/

over 20,000 academic journals: http://zetoc.mimas.ac.uk/index.html

well over a million articles: http://informationr.net/ir/14-1/
paper391.html

Nullius in Verba

recent study hunting down: http://www.badscience.net/2010/06/
2008/01/washing-the-numbers-selling-the-model/

strengths and weaknesses of the HES dataset: http://www.health-
knowledge.org.uk/parta/paper1knowledge/3_healthinformation/3a_
Populations/3a2.asp

standard public health exam: http://www.publichealthy.com/partatips.
htm

couldn't find the pattern either: http://www.guardian.co.uk/commen-
tisfree/2010/jun/14/size-isnt-everything-in-hospital-treatment

Is It OK to Ignore Results from People You Don't Trust?

Is it OK: http://www.badscience.net/2010/03/when-is-it-okay-to-
ignore-people-you-dont-trust/

three month period in 2002: http://www.kingsfund.org.uk/research/
publications/health_in_the.html

systematic review on the subject: http://iospress.metapress.com/
content/x880352113361jk4/
Legacy Tobacco Documents Library: http://legacy.library.ucsf.edu/
The importance of younger adults: http://legacy.library.ucsf.edu:8080/
e/y/n/eyn18c00/Seyn18c00.pdf
Youth cigarette – new concepts: http://legacy.library.ucsf.edu:8080/
o/y/q/oyq83f00/Soyq83f00.pdf
Two researchers, Schairer and Schöniger: http://ije.oxfordjournals.
org/cgi/content/full/30/1/31

Foreign Substances in Your Precious Bodily Fluids
Foreign Substances: http://www.badscience.net/2008/02/foreign-
substances-in-your-precious-bodily-fluids/
calls it a 'poison': http://www.sunderlandecho.com/news/Fluoride-
39poison39-says-Wear-MP.3745368.jp
lead and arsenic: http://www.guardian.co.uk/weekend/story/0,,1
556071,00.html
speaking way beyond the evidence: http://www.bmj.com/cgi/
content/full/335/7622/699
the benefits of fluoridation: http://www.york.ac.uk/inst/crd/fluorid.
htm
'overwhelming evidence': http://www.thisislondon.co.uk/news/
article-23435556-details/Fluoride+to+be+added+to+all+Britain
%27s+tap+water+to+tackle+tooth+decay/article.do
Alan Johnston says: http://ukpress.google.com/article/
ALeqM5iJ2QJhA7EJ_HT5WiESAO3L1bla6Q
people are misrepresenting: http://www.york.ac.uk/inst/crd/fluoridnew.
htm
Further references:
The most readable overview on the poor quality of the data is this
article by Cheng, Chalmers and Sheldon. Prof Sir Iain Chalmers
founded the Cochrane library, for anyone who seeks to doubt
his badassness: www.bmj.com/cgi/content/full/335/7622/699
The Taiwan study is here (the authors attribute the increase in risk
to a problem of multiple comparisons): http://www.science
direct.com/science/article/pii/S0013935199940185
The York review is here: www.york.ac.uk/inst/crd/fluorid.htm. You'll
see that for several years now they've been trying to point out that

people are misrepresenting their work: http://www.york.ac.uk/inst/
crd/fluoridnew.htm

And if you want a particularly good example of someone pomp-
ously overstating the evidence, try this doozy from the
Sunday Mirror: www.sundaymirror.co.uk/news/columnists/
opinion/2008/02/03/fluoride-the-whole-
tooth-98487-20307072/

How Myths Are Made

How Myths are Made: http://www.badscience.net/2009/08/how-
myths-are-made/

successfully used a court order: http://speakingofmedicine.plos.
org/2009/08/05/guest-blog-from-adriane-fugh-berman-plos-
medicine-and-the-new-york-times-victorious-in-court-public-
will-have-access-to-ghostwriting-documents/

Such an analysis: http://www.bmj.com/cgi/content/abstract/339/
jul20_3/b2680

Publish or Be Damned

Publish or be Damned: http://www.badscience.net/2005/08/publish-
or-be-damned/

Academic Papers Are Hidden from the Public. Here's Some Direct Action

Academic Papers are Hidden: http://www.badscience.net/2011/09/
academic-papers-are-hidden-from-the-public-heres-some-direct-
action/

piece on academic publishers: http://www.guardian.co.uk/comment-
isfree/2011/aug/29/academic-publishers-murdoch-socialist

pay up front: http://www.timeshighereducation.co.uk/story.asp?section
code=26&storycode=417266&c=1

Aaron Swartz: http://blog.demandprogress.org/2011/07/federal-
government-indicts-former-demand-progress-executive-director-
for-downloading-too-many-journal-articles/

federal indictment document online: http://web.mit.edu/bitbucket/
Swartz,Aaron Indictment.pdf

harvest academic papers from JSTOR: http://about.jstor.org/news-
events/news/jstor-statement-misuse-incident-and-criminal-case

Pirate Bay: http://thepiratebay.org/torrent/6554331/Papers_from_
    Philosophical_Transactions_of_the_Royal_Society__fro
Royal Society Papers: http://royalsociety.org/about-us/reporting/

BIOLOGISING

Neuro-Realism

Neuro-Realism: http://www.badscience.net/2010/10/neuro-realism/
When the BBC tells you: http://www.bbc.co.uk/news/health-
    11620971
'normal' sex drive http://amzn.to/doM9h8
'hypoactive sexual desire disorder': http://en.wikipedia.org/wiki/
    Hypoactive_sexual_desire_disorder
tells the *Mail*: http://www.dailymail.co.uk/news/article-1323730/
    Is-women-dont-like-make-love-Scientists-discover-low-libidos-
    behave-differently.html
In the *Metro*: http://www.metro.co.uk/lifestyle/845156-women-with-
    low-sex-drive-have-different-brains
'fMRI in the Public Eye': http://www.ncbi.nlm.nih.gov/pmc/articles/
    PMC1524852/

The Stigma Game

The Stigma Game: http://www.badscience.net/2010/10/pride-and-
    prejudice/
looked for chromosomal deletions: http://dx.doi.org/10.1016/
    S0140-6736%2810%2961109-9
including that in the *Guardian*: http://www.guardian.co.uk/
    society/2010/sep/30/hyperactive-children-genetic-disorder-
    study
said Professor Anita Thapar: http://www.guardian.co.uk/society/
    2010/sep/30/hyperactive-children-genetic-disorder-study
Read and Harre: http://informahealthcare.com/doi/
    abs/10.1080/09638230123129
Read and Law: http://isp.sagepub.com/content/45/3/216.full.
    pdf+html
Walker and Read: http://www.ncbi.nlm.nih.gov/pubmed/12530335
Dietrich and colleagues: http://onlinelibrary.wiley.com/doi/10.1111/
    j.1440-1614.2004.01363.x/abstract

review of the literature to date: http://onlinelibrary.wiley.com/doi/
10.1111/j.1600-0447.2006.00824.x/full
'Genetic Bases of Mental Illness': http://dx.doi.org/10.1016/
S0166-2236%2802%2902209-9

Pink, Pink, Pink, Pink. Pink Moan

Pink Moan: http://www.badscience.net/2007/08/pink-pink-pink-pink-
pink-moan/
every single newspaper in the world: http://news.google.co.uk/
news?ie=UTF-8&oe=UTF-8&rls=org.
said *The Times*: http://www.timesonline.co.uk/tol/news/uk/science/
article2294539.ece
The study took 208: http://dx.doi.org/10.1016/j.cub.2007.06.022
Bem Sex Role Inventory: http://www.neiu.edu/~tschuepf/bsri.
html
Anyone can take this test online: http://garote.bdmonkeys.net/bsri.
html
Further references:
The academic paper is here: Anya C. Hurlbert and Yazhu Ling,
'Biological components of sex differences in color preference',
Current Biology, Volume 17, Issue 16, 21 August 2007, Pages
R623–R625, dx.doi.org/10.1016/j.cub.2007.06.022

STATISTICS

Guns Don't Kill People, Puppies Do

Guns Don't Kill People: http://www.badscience.net/2010/02/guns-
dont-kill-people-puppies-do/
3 babies on same date: http://www.express.co.uk/posts/view/157444/
Mum-beats-odds-of-50m-to-one-to-have-3-babies-on-same-
date-
725,440 births: http://www.statistics.gov.uk/cci/nugget.asp?id=369
'Who is Having Babies': http://www.statistics.gov.uk/pdfdir/births
1209.pdf
'Dog Shoots Man': http://www.cbsnews.com/stories/2010/02/01/ap/
strange/main6162474.shtml
'Dog Shoots Man in Back': http://www.foxnews.com/story/
0,2933,291687,00.html

another in Iowa: http://news.bbc.co.uk/1/hi/world/americas/7068549.
stm

Puppy Shoots Man: http://www.msnbc.msn.com/id/5950304

Datamining for Terrorists Would Be Lovely If It Worked
Datamining for Terrorists: http://www.badscience.net/2009/02/data
mining-would-be-lovely-if-it-worked/

This week Sir David Omand: http://www.guardian.co.uk/uk/2009/
feb/25/database-state-ippr-paper

described how the state: http://www.ippr.org/publicationsandreports/
publication.asp?id=646

what statisticians call the 'specificity': http://en.wikipedia.org/wiki/
Sensitivity_(tests)

Benford's Law: Using Stats to Bust an Entire Nation for Naughtiness
Benford's Law: http://www.badscience.net/2011/09/benfords-
law-using-stats-to-bust-an-entire-nation-for-naughtiness/

something called Benford's Law: http://en.wikipedia.org/wiki/
Benford%27s_law

61,838,154 in 2009: http://www.google.co.uk/publicdata/
explore?ds=d5bncppjof8f9_&met_y=sp_pop_totl&idim=country:
GBR&dl=en&hl=en&q=uk+population

think about why this happens: http://en.wikipedia.org/wiki/
Benfords_law

testingbenfordslaw.com: http://testingbenfordslaw.com/

Twitter users' follower counts: http://testingbenfordslaw.com/twitter-
users-by-followers-count

books in different libraries: http://testingbenfordslaw.com/total-
number-of-print-materials-in-us-libraries

countries' economic data: http://econpapers.repec.org/article/
blagermec/v_3a10_3ay_3a2009_3ai_3a_3ap_3a339-351.htm

the results were published: http://onlinelibrary.wiley.com/
doi/10.1111/j.1468-0475.2011.00542.x/abstract

hat-tip to Tim Harford: http://timharford.com/2011/09/look-out-for-
no-1/

macroeconomic data: http://epp.eurostat.ec.europa.eu/portal/page/
portal/esa95_supply_use_input_tables/data/workbooks

online repository Eurostat: http://epp.eurostat.ec.europa.eu/portal/
  page/portal/esa95_supply_use_input_tables/introduction
run several investigations: http://epp.eurostat.ec.europa.eu/portal/
  page/portal/product_details/publication?p_product_code=COM_
  2010_report_greek
repeat the analysis for yourself: http://rstudio.org/

The Certainty of Chance
The Certainty of Chance: http://www.badscience.net/2008/09/the-
  certainty-of-chance/

**Sampling Error, the Unspoken Issue Behind Small Number
Changes in the News**
Sampling Error: http://www.badscience.net/2011/08/untitled-1/
"Worrying" Jobless Rise: http://www.bbc.co.uk/news/uk-politics-
  14558369
'Labour Market' figures: http://www.statistics.gov.uk/StatBase/
  Product.asp?vlnk=1944
in a PDF document: http://www.statistics.gov.uk/pdfdir/lmsuk0811.pdf

**Scientific Proof That We Live in a Warmer and More Caring
Universe**
Scientific Proof: http://www.badscience.net/2008/11/scientific-proof-
  that-we-live-in-a-warmer-and-more-caring-universe/
'Down's births increase in a caring Britain': http://www.timeson
  line.co.uk/tol/life_and_style/health/article5219174.ece
when diagnosed with Down's: http://www.dailymail.co.uk/news/
  article-1088774/More-mothers-choosing-babies-diagnosed-Downs-
  Syndrome.html
said the *Independent*: http://www.independent.co.uk/life-style/health-
  and-wellbeing/health-news/downs-parents-think-again-1032193.
  html
said the *Mirror*: http://www.mirror.co.uk/news/top-stories/2008/11/24/
  babies-born-with-down-s-syndrome-on-increase-115875-20919849/
Radio 4 documentary: http://www.bbc.co.uk/programmes/b00fkx0m
her what plays Ruth Archer: http://www.bbc.co.uk/radio4/archers/
  whos_who/actors/actor_felicity_finch.shtml

National Down Syndrome Cytogenetic Register: http://www.wolfson.
  qmul.ac.uk/ndscr/
took the story to pieces: http://www.wolfson.qmul.ac.uk/ndscr/
National Down Syndrome Cytogenetic Register: http://www.nhs.uk/
  news/2008/11November/Pages/DownssyndromeQA.aspx
Radio 4 website: http://www.bbc.co.uk/radio4/

Drink Coffee, See Dead People
Drink Coffee: http://www.badscience.net/2009/01/drink-coffee-see-
  dead-people/
7 cups of coffee a day: http://www.express.co.uk/posts/view/79820/
  utter-cock-as-usual
in almost every national newspaper: http://news.google.co.uk/news?
  q=coffee+hallucinations+location%3Auk
exactly what the researchers did: http://dx.doi.org/10.1016/j.paid.
  2008.10.032
survey is still online: http://psychology.dur.ac.uk:82/srj/caffeine2.
  html
'Launay-Slade Hallucination Scale': http://www.google.co.uk/
  search?q=Launay-Slade+Hallucination+Scale
alternative explanations: http://en.wikipedia.org/wiki/Confounding_
  variable
the academic paper: http://dx.doi.org/10.1016/j.paid.2008.10.032
the press release: http://www.alphagalileo.org/index.cfm?fuseaction=
  readrelease&releaseid=535120&ez_search=1
'multiple comparisons': http://en.wikipedia.org/wiki/Multiple_
  comparisons
no one was there: http://www.thinkgeek.com/caffeine/accessories/
  5a65/
draw a target around them: http://en.wikipedia.org/wiki/Texas_
  sharpshooter_fallacy

Voices of the Ancients
Voices of the Ancients: http://www.badscience.net/2010/01/voices-
  of-the-ancients/
Daily Mail: http://www.dailymail.co.uk/sciencetech/article-1240746/
  Prehistoric-sat-nav-set-ancestors-Britain.html

the *Metro*: http://www.metro.co.uk/news/807855-did-prehistoric-
satnav-help-britons-find-their-way
Matt Parker: http://www.standupmaths.com/
applied the same techniques: http://bengoldacre.posterous.com/did-
aliens-play-a-role-in-woolworths

BIG DATA

There's Something Magical About Watching Patterns Emerge from Data
There's Something Magical: http://www.badscience.net/2011/06/
theres-something-magical-about-watching-patterns-emerge-from-
data/
British Medical Journal: http://www.bmj.com/content/342/bmj.d2983.
full
first NHS reforms: http://www.guardian.co.uk/society/nhs

Give Us the Data
A consultation is under way: http://c561635.r35.cf2.rackcdn.com/A-
Consultation-on-Data-Policy-for-a-Public-Data-Corporation.pdf
foolishly restrictive: http://pdcconsult.ernestmarples.com/
everyday government data: http://www.theguardian.com/politics/
government-data
forbidden to repurpose it: http://hadleybeeman.net/2011/01/26/uses-
for-open-data/
TheyWorkForYou.com: http://www.theyworkforyou.com/
all our postcode information: http://en.wikipedia.org/wiki/Postcodes_
in_the_United_Kingdom
the house-number boundaries: http://ernestmarples.com/blog/
2010/01/postcode-petition-response-our-reply/
make the government: http://pdcconsult.ernestmarples.com/

Care.data Can Save Lives: But Not If We Bungle It
greatest need in the NHS http://www.theguardian.com/society/nhs
at risk by the bungled: http://www.nature.com/news/power-to-the-
people-1.14505?WT.ec_id=NATURE-20140116
implementation of the care.data: http://www.bbc.co.uk/news/health-
26187980

on hold for six months: http://www.theguardian.com/society/2014/
feb/18/nhs-delays-sharing-medical-records-care-data
by and large, the public: http://www.wellcome.ac.uk/About-us/
Publications/Reports/Public-engagement/WTP053206.htm
results of clinical trials: http://www.badscience.net/category/
publication-bias/
Tim Kelsey: http://www.theguardian.com/healthcare-network/2013/
aug/21/tim-kelsey-nhs-big-data
Dr Foster Intelligence http://drfosterintelligence.co.uk/solutions/
nhs-hospitals/
announcing boldly: http://www.bbc.co.uk/news/uk-england-devon-2
6030479
identifiable patient data: http://www.out-law.com/en/arti-
cles/2014/february/pseudonymised-health-data-will-not-be-
able-to-be-traced-back-to-individuals-under-caredata-scheme-
says-official-/
medical research database: http://www.theguardian.com/science/
medical-research
keen to point out: http://www.ehi.co.uk/news/ehi/9207/re-identifying-
care.data-illegal
Steve Tennison: http://ico.org.uk/news/latest_news/2013/
gp-surgery-manager-prosecuted-for-illegally-accessing-patients-
medical-records-02122013

Care.data Has Been Bungled

millions of patients: http://www.telegraph.co.uk/health/nhs/10659147/
Patient-records-should-not-have-been-sold-NHS-admits.html
information governance assessment: http://www.hscic.gov.uk/media/
12866/caredata-addendum---Information-Governance-Assessment/
pdf/care.data_addendum_-_IG_assessment_-_September_2013_
(NIC-178106-MLSWX.A0913).pdf

SURVEYS

The Huff

The Huff: http://www.badscience.net/2008/01/the-huff/
remains in print: How to Lie: http://www.amazon.co.uk/How-Lie-
Statistics-Penguin-Business/dp/0140136290/tag=bs0b-21

Doctors say no to abortions: http://www.telegraph.co.uk/news/
    main.jhtml?xml=/news/2007/12/29/nabort129.xml
I hope that DNUK: http://www.doctors.net.uk/

A New and Interesting Form of Wrong
A New and Interesting: http://www.badscience.net/2010/11/1864/
reported in the *Guardian*: http://www.guardian.co.uk/world/2010/
    nov/15/ gay-people-coming-out-younger-age
its press release: http://www.stonewall.org.uk/media/current_releases/
    4867.asp
an interesting problem: http://bengoldacre.posterous.com/an-
    interesting-survival-analysis-problem-via

'Hello Madam, Would You Like Your Children to Be Unemployed?'
Hello Madam: http://www.badscience.net/2010/11/hello-madam-
    would-you-like-your-children-to-be-unemployed/
secret nuclear bunker: http://www.secretnuclearbunker.com/
Project Redsand: http://www.project-redsand.com/
travelled to Dungeness: http://en.wikipedia.org/wiki/Dungeness_%28
    headland%29
toytown narrow-gauge railway: http://www.rhdr.org.uk/
Derek Jarman's house: http://www.guardian.co.uk/lifeandstyle/
    allotment/2008/ feb/27/dereksretreat
terrifying nuclear power station: http://en.wikipedia.org/wiki/
    Dungeness_Nuclear_Power_Station
BBC dutifully reported: http://www.bbc.co.uk/news/uk-england-
    somerset-11521839
the original polling questions: http://www.climatesock.com/2010/11/
    don%E2%80%99t-just-believe-what-you%E2%80%99re-told-
    about-polls/

EPIDEMIOLOGY

Beau Funnel
The BBC has found a story: UK bowel cancer rates: http://www.
    theguardian.com/media/bbc

variation in UK bowel cancer rates: http://www.bbc.co.uk/news/
health-14854019

average death rate: http://www.theguardian.com/society/bowel-
cancer

Paul Barden: http://pb204.blogspot.com/

decided to download the data: http://pb204.blogspot.com/2011/09/
three-fold-variation-in-uk-bowel-cancer.html

come from a press release: http://www.beatingbowelcancer.org/
news/sep2011/charity-warns-wide-variations-bowel-cancer-
death-rates-must-not-be-ignored

Beating Bowel Cancer: http://www.theguardian.com/society/cancer

built a map: http://www.bowelcancermap.org/

the Poisson distribution: http://en.wikipedia.org/wiki/Poisson_
distribution

bell-shaped curve: http://www.etsy.com/listing/48582479/standard-
normal-distribution-plushie

series of simulations: http://pb204.blogspot.com/2011/09/three-fold-
variation-in-uk-bowel-cancer.html

Understanding Uncertainty: http://understandinguncertainty.org/

Spiegelhalter suggested that Barden: http://pb204.blogspot.com/
2011/09/im-grateful-to-david-spiegelhalter-of.html

Comparing Institutional Performance: http://medicine.cf.ac.uk/media/
filer_public/2010/10/11/journal_club_-_spiegelhalter_stats_in_med_
funnel_plots.pdf

the citation classic: http://scholar.google.co.uk/scholar?cluster=
12057401031362296814

The Public Health Observatories: http://www.apho.org.uk/

several neat tools: http://www.apho.org.uk/default.aspx?QN=HP_
INTERACTIVE2011

draw a funnel plot: http://tools.erpho.org.uk/poisson.aspx

When Journalists Do Primary Research

When Journalists: http://www.badscience.net/2011/04/when-
journalists-do-primary-research/

said the *Telegraph*: http://www.telegraph.co.uk/health/8434106/
Recession-linked-to-huge-rise-in-use-of-antidepressants.
html

said the *Daily Mail*: http://www.dailymail.co.uk/health/article-
1374284/Depression-Economic-slump-fuels-43-rise-use-anti-
depressants.html?ito=feeds-newsxml

being handed antidepressants: http://www.express.co.uk/posts/
view/239413/Money-worries-driving-more-to-pills

the *Guardian* joined in: http://www.guardian.co.uk/society/2011/
apr/07/dramatic-rise-antidpressant-prescriptions-money-
worries

seems to have come from BBC: http://www.bbc.co.uk/news/health-
12986314

2009 data: http://www.ic.nhs.uk/statistics-and-data-collections/primary-
care/prescriptions/prescription-cost-analysis-england-- 2009

due about now: http://www.ic.nhs.uk/statistics-and-data-collections/
primary-care/prescriptions

antidepressant prescribing: http://www.bmj.com/content/339/bmj.
b3999.full

British Journal of General Practice: http://www.ncbi.nlm.nih.gov/
pmc/articles/PMC1839016/ ?tool=pubmed

2.8 million: http://2.8.mil/

The *BMJ* paper: http://www.bmj.com/content/339/bmj.b3999.full

Confound You!

Unplanned children develop more slowly: http://www.foxnews.
com/health/2011/07/27/unplanned-children-develop-more-
slowly/

IVF children have bigger vocabulary: http://www.telegraph.co.uk/
education/educationnews/8663105/IVF-children-have-bigger-
vocabulary-than-unplanned-babies.html

Children born after an unwanted pregnancy: http://www.eurekalert.
org/pub_releases/2011-07/ bmj-cba072511.php

this *BMJ* paper: http://www.bmj.com/content/343/bmj.d4473.full

Bicycle Helmets and the Law

Dennis and colleagues: http://www.bmj.com/content/346/bmj.
f2674

'seems to have been minimal': http://www.bmj.com/content/346/
bmj.f2674?ijkey=f5e18ac0289812fef1140ebb4f839638e47d085
c&keytype2=tf_ipsecsha

come to different conclusions: http://www.ncbi.nlm.nih.gov/
pubmed/18646128?access_num=18646128&link_type=MED&dopt
=Abstract

less likely to have a head injury: http://summaries.cochrane.org/
CD001855/INJ_wearing-a-helmet-dramatically-reduces-the-risk-
of-head-and-facial-injuries-for-bicyclists-involved-in-a-crash-
even-if-it-involves-a-motor-vehicle

'documented in many fields: http://onlinelibrary.wiley.com/
doi/10.1111/j.1539-6924.2011.01589.x/abstract and Adams, J.,
Risk. Taylor & Francis, 2002.

larger clearance to cyclists without: http://www.ncbi.nlm.nih.gov/
pubmed/17064655?access_num=17064655&link_type=MED&dopt
=Abstract

outweigh the risks of crashes: http://www.euro.who.int/en/health-
topics/disease-prevention/physical-activity/activities/hepa-europe/
hepa-europe-projects-and-working-groups/development-of-
methods-for-quantification-of-health-benefits-from-walking-and-
cycling

such as for children: http://onlinelibrary.wiley.com/doi/10.1111/
j.1539-6924.2011.01785.x/abstract

cycling in the second group: http://www.sciencedirect.com/science/
article/pii/S1369847812000587

Smeed's Law: http://www.ncbi.nlm.nih.gov/pubmed/16389930?access_
num=16389930&link_type=MED&dopt=Abstract

speculative assumptions: http://onlinelibrary.wiley.com/doi/10.1111/
j.1539-6924.2012.01770.x/abstract

emotional response: http://psycnet.apa.org/?&fa=main.doiLanding&
doi=10.1037/0278-6133.24.4.S35

Screen Test

Screen Test: http://www.badscience.net/2008/01/screen-test/

Screening and repairing: http://www.mrw.interscience.wiley.com/
cochrane/clsysrev/articles/CD002945/frame.html

mammogram screening for breast cancer: http://www.pubmedcentral.
nih.gov/articlerender.fcgi?artid=556337

Researchers have studied: http://www.bmj.com/cgi/content/
full/332/7540/538

repeatedly been shown: http://www.bmj.com/cgi/content/abstract/3
28/7432/148?ijkey=c30d4caac1758e2a62478825d4943f10820ae34c
&keytype2=tf_ipsecsha
at least one large survey: http://jama.ama-assn.org/cgi/content/
abstract/291/1/71?ijkey=5b96d0b882f4bef1847b1dc3364754c3bbd
31b05&keytype2=tf_ipsecsha

How Do You Know?
mobile phones: http://www.theguardian.com/technology/
mobilephones
cause brain cancer: http://www.guardian.co.uk/science/2011/
may/31/mobile-phone-radiation-cancer-risk
part of the WHO: http://en.wikipedia.org/wiki/International_Agency_
for_Research_on_Cancer
triggered over 3000 articles: http://en.wikipedia.org/wiki/World_
Health_Organisation
only a press release: http://news.google.co.uk/news/story?pz=1&cf
=all&ned=uk&hl=en&q=iarc&ncl=dZVLtSHEkU8Cq_
MxKk 87AKCKGvSrM
limits of the research: http://www.iarc.fr/en/media-centre/pr/2011/
pdfs/pr208_E.pdf
hasn't changed much: http://journals.lww.com/epidem/Fulltext/
2009/09000/Health_Effects_of_Mobile_Telephones.6.aspx
in every 100,000: http://www.statistics.gov.uk/StatBase/Product.
asp?vlnk=7720
'prospective cohort study': http://www.statistics.gov.uk/StatBase/
Product.asp?vlnk=7720
'prospective cohort study': http://en.wikipedia.org/wiki/Cohort_study
'retrospective case-control study': http://resources.bmj.com/bmj/
readers/readers/epidemiology-for-the-uninitiated/8-case-control-
and-cross-sectional-studies
Interphone study: http://ije.oxfordjournals.org/content/39/3/675.
abstract

Anecdotes Are Great. If They Really Illustrate the Data
Channel 4 News: http://www.channel4.com/news/catch-up/display/
playlistref/250711/clipid/250711_4ON_DUCHENNE_25

Study shows no such thing: http://www.thelancet.com/journals/
lancet/article/PIIS0140-6736%2811%2960756-3/abstract
Great Ormond Street press release: http://bgarchive.posterous.com/
first-targeted-treatment-success-for-duchenne
tracked down online: http://twitter.com/%C2%A3%21/uclnews/
status/95506900220776449
Evan Harris: http://www.guardian.co.uk/science/political-science
has been supoptimal: http://www.badscience.net/2009/05/dodgy-
academic-pr/
biggest diseases in medicine: http://www.theguardian.com/education/
medicine
patients on doxazosin: http://ebm.bmj.com/content/5/6/172.full.
pdf
blood test called HbA1c: http://en.wikipedia.org/wiki/Glycated_
hemoglobin
reduce your HbA1c level: http://www.bmj.com/content/341/bmj.
c4805.full
turned out that rosiglitazone: http://www.badscience.net/index.
php?s=rosiglitazone
drug has now been suspended: http://www.ema.europa.eu/ema/index.
jsp?curl=pages/news_and_events/news/2010/09/news_detail_
001119. jsp&murl=menus/news_and_events/news_and_events.
jsp&mi d=WC0b01ac058004d5c1&jsenabled=false

The Strange Case of the Magnetic Wine
The Strange Case: http://www.badscience.net/2003/12/the-strange-
case-of-the-magnetic-wine/

What Is Science?: First, Magnetise Your Wine
What Is Science: http://www.badscience.net/2005/12/what-is-
science-first-magnetise-your-wine/

BAD ACADEMIA

What If Academics Were as Dumb as Quacks with Statistics
What if Academics: http://www.badscience.net/2011/10/what-if-
academics-were-as-dumb-as-quacks-with-statistics/

publish a mighty torpedo: http://www.nature.com/neuro/journal/
v14/n9/full/nn.2886.html

**Brain-Imaging Studies Report More Positive Findings Than
Their Numbers Can Support. This Is Fishy**
Brain-Imaging Studies: http://www.badscience.net/2011/08/brain-
imaging-studies-report-more-positive-findings-than-their-numbers-
can-support-this-is-fishy/
publication bias:http://www.badscience.net/category/publication-bias/
took a different approach: http://archpsyc.ama-assn.org/cgi/
content/abstract/archgenpsychiatry.2011.28

'None of Your Damn Business'
None of Your: http://www.badscience.net/2011/01/none-of-your-
damn-business/
2004 published a study: http://ats.ctsnetjournals.org/cgi/content/
abstract/ annts;78/4/1433
it was retracted: http://retractionwatch.wordpress.com/2011/01/04/
thoracic-surgery-journal-retracts-hypertension-study-marred-by-
troubled-data/
Dr L. Henry Edmunds Jr, MD: http://www.uphs.upenn.edu/
surgery/faculty/lhe.html
none of your damn business: http://retractionwatch.wordpress.
com/2011/01/05/why-was-that-paper-retracted-editor-to-retraction
-watch-its-none-of-your-damn-business/
retracted a 2009 paper: http://retractionwatch.wordpress.com/
2010/12/17/journal-of-the-american-chemical-society-retracts-
gold-nanoparticle-paper/
ad hoc blog tracking: http://retractionwatch.wordpress.
com/2010/08/03/why-write-a-blog-about-retractions/
stories behind retractions: http://retractionwatch.wordpress.
com/2010/10/22/update-on-axel-ullrich-retractions-lead-author-
manipulated-figures-says-ullrich/
study last year: http://informahealthcare.com/doi/
abs/10.1185/03007991003603804
policing academic misconduct: http://www.bmj.com/
content/331/7511/288.full

Twelve Monkeys. No . . . Eight. Wait, Sorry, I Meant Fourteen
Twelve Monkeys: http://www.badscience.net/2010/01/12-monkeys-
no-8-wait-sorry-i-meant-14/
Animals in Research: http://www.nc3rs.org.uk/
PLoS One: http://www.plosone.org/article/info:doi/10.1371/journal.
pone.0007824

Medical Hypotheses Fails the Aids Test
Medical Hypotheses: http://www.badscience.net/2009/09/medical-
hypotheses-fails-the-aids-test/
in the newspapers: http://news.google.co.uk/news/more?um=1&ned
=uk&cf=all&ncl=dEYVi6Gb688Hm1MSCLQllCltTV9pM
Peer review: http://en.wikipedia.org/wiki/Peer_review
Aidstruth.org: http://aidstruth.org/about
Elsevier have withdrawn: http://aidstruth.org/news/2009/elsevier-
retracts-duesberg%E2%80%99s-aids-denialist-article
one surreally crass paper: http://www.badscience.net/2007/08/
observations-on-the-classification-of-idiots/
Italian doctors argued: http://www.badscience.net/2007/08/
observations-on-the-classification-of-idiots/
treatment for nasal congestion: http://www.badscience.net/2008/10/
more-crap-journals/
Peter Duesberg and David Rasnick: http://www.ncbi.nlm.nih.gov/
pubmed/19619953
Lancet paper they reference: http://www.ncbi.nlm.nih.gov/pubmed/
16890831
Bruce Charlton has argued: http://www.bmj.com/cgi/content/
extract/335/7617/451
written to Medline: http://www.aidstruth.org/sites/aidstruth.org/
files/NLMLetter-2009.08.05.pdf
deaths of an estimated: http://afraf.oxfordjournals.org/cgi/content/
abstract/107/427/157
330,000 people: http://www.ncbi.nlm.nih.gov/pubmed/18931626

Observations on the Classification of Idiots
Observations on: http://www.badscience.net/2007/08/observations-
on-the-classification-of-idiots/

More Crap Journals?
More Crap Journals: http://www.badscience.net/2008/10/more-crap-journals/
potential treatment of nasal congestion: http://www.medical-hypotheses.com/article/S0306- 9877(08)00115-1/fulltext
unreliable and potentially hazardous: http://www.sciencedirect.com/science/article/pii/S030698770800354X
surreally crass paper: http://www.badscience.net/2007/08/observations-on-the-classification-of-idiots/
more from him online: http://www.bmj.com/cgi/content/extract/335/7617/451
only one in four: http://www.badscience.net/2008/09/seriously-is-the-daily-mail-any-worse-than-your-average-academic-journal/

GOVERNMENT STATISTICS

If You Want to Be Trusted More: Claim Less
If You Want: http://www.badscience.net/2010/01/if-you-want-to-be-trusted-more-claim-less/
Sunday Times: http://business.timesonline.co.uk/tol/business/industry_sectors/public_sector/article6974029.ece
copycat story: http://www.telegraph.co.uk/finance/6925897/Public-pay-races-ahead-in-recession.html
Survey of Hours and Earnings: http://www.statistics.gov.uk/statBase/product.asp?vlnk=13101
ASHE 2009 data: http://www.statistics.gov.uk/downloads/theme_labour/ASHE-2009/tab1_5a.xls

Is This the Worst Government Statistic Ever Created?
costs every household £452: http://www.dailymail.co.uk/news/article-2004590/Council-incompetence-costing-household-Britain-452-year.html?ito=feeds-newsxml
the *Express* agreed: http://www.express.co.uk/posts/view/253235/Councils-owe-you-rebate
proper story, from press release: http://www.communities.gov.uk/news/corporate/1925280
Department of Communities and Local Government: http://www.guardian.co.uk/society/localgovernment

Notes

'Opera Solutions White Paper': http://www.badscience.net/
wp-content/uploads/Opera-gov-savings-local-gov.pdf
The 'full report': http://www.badscience.net/wp-content/uploads/
Opera-gov-savings-local-gov.pdf
the biggest thing: http://www.niepbuiltenvironment.org.uk/
documents/ReportonPotentialLocalGovernmentSavingsthrough
BetterProcurementF.doc

Anarchy for the UK. Ish.
Anarchy for the UK: http://www.badscience.net/2011/04/anarchy-
for-the-uk-ish/
The *Sun* said: http://www.thesun.co.uk/sol/homepage/
news/3494359/Cops-charge-149-after-protest-against-cuts-turns-
ugly.html
149 people charged: http://www.guardian.co.uk/society/2011/
mar/28/cuts-protest-violence-149-chargedhttp://www.guardian.
co.uk/society/2011/mar/28/cuts-protest-violence-149-charged
the *Manchester Evening News* carried: http://menmedia.co.uk/
manchestereveningnews/news/s/1416475_boy-17-from-manchester-
among-149-charged-over-violence-after-anti-cuts-march
138 were people: http://www.met.police.uk/pressbureau/Bur27/
page01.htm
peaceful occupation of Fortnum & Mason: http://www.guardian.
co.uk/uk/2011/mar/28/cuts-protest-uk-uncut-fortnum
systematic review: http://www.bmj.com/content/340/bmj.c2369.full
sports participation in Barcelona: http://olympicstudies.uab.es/pdf/
wp039_eng.pdf
being quietly dropped: http://www.guardian.co.uk/sport/2011/
mar/28/jeremy-hunt-london-2012-legacy

More Than Sixty Children Saved from Abuse
More than Sixty: http://www.badscience.net/2010/08/more-than-
60-children-saved-from-abuse/
during its pilot: http://www.homeoffice.gov.uk/media-centre/news/
child-protection-scheme
the *Sun* said: http://www.thesun.co.uk/sol/homepage/
news/3077347/Sarahs-Law-hailed-a-success-as-scheme-rolls-out-
nationwide.html

an excellent report: http://rds.homeoffice.gov.uk/rds/pdfs10/horr
32c.pdf

Conrad Quilty-Harper in the *Telegraph*: http://blogs.telegraph.co.
uk/news/conradquiltyharper/ 100049409/the-sarahs-law-trial-
didnt-save-60-children-from-sex-offenders/

FullFact: http://www.fullfact.org/blogdetail/sarahs_law_the_story_
behind_the_statistics

Home Taping Didn't Kill Music

Home Taping: http://www.badscience.net/2009/06/home-taping-
didnt-kill-music/

Costs Billions: http://www.thesun.co.uk/sol/homepage/
news/2454908/Downloading-costs-billions.html

Daily Mail was worried: http://www.dailymail.co.uk/sciencetech/
article-1189509/Illegal-file-sharing-downloads-cost-thousands-
British-jobs-year.htm

also buy more music: http://www.guardian.co.uk/music/2009/
apr/21/study-finds-pirates-buy-more-music

the original report: http://www.sabip.org.uk/sabip-ciberreport.pdf

called CIBER: http://www.ucl.ac.uk/infostudies/research/ciber/

called SABIP: http://www.sabip.org.uk/

full CIBER documents: http://www.ucl.ac.uk/infostudies/research/
ciber/

2004 press release: http://www.iprights.com/publications/Alert_156.pdf

by a BBC journalist: http://news.bbc.co.uk/1/hi/technology/
8073068.stm

Is This a Joke?

Is This a Joke: http://www.badscience.net/2009/07/is-this-a-joke/

published a consultation paper: http://www.homeoffice.gov.uk/
documents/cons-2009-dna-database/

EVIDENCE-BASED POLICY

I'd Expect This from UKIP or the *Daily Mail*. Not from a Government Leaflet

I'd Expect This: http://www.badscience.net/2011/04/id-expect-this-
from-ukip-or-the-daily-mail-not-from-a-government-leaflet/

For A Stronger NHS: http://www.dh.gov.uk/prod_consum_dh/
groups/dh_digitalassets/documents/digitalasset/dh_125855.pdf
paper in the *British Journal of Cancer*: http://www.nature.com/bjc/
journal/v101/n2s/full/6605401a.html
we've seen this a lot: http://www.badscience.net/2011/04/2011/02/
why-is-evidence-so-hard-for-politicians/
NHS workforce data: http://www.ic.nhs.uk/webfiles/publications/
010_Workforce/provisionalmonthlyhchsworkforce/Dec10/
FINAL_Table_ National.xls
NHS Information Centre figures: http://www.ic.nhs.uk/pubs/
nhsworkforce
total number of doctors: http://www.ic.nhs.uk/webfiles/publications/
010_Workforce/nhsstaff9909/NHS_Staff_1999_2009_Master_
Table.xls
British Social Attitudes Survey: http://www.straightstatistics.org/
blog/2011/03/29/kings-fund-rescue-social-attitudes-survey
Question 583: http://www.natcen.ac.uk/media/299809/bsa2007
questionnaire. pdf
that costs £52: http://www.uk.sagepub.com/booksProdDesc.
nav?prodId=Book 233725
defies all reason: http://www.guardian.co.uk/commentisfree/2011/
apr/13/andrewlansley-health

Andrew Lansley and His Imaginary Evidence

Andrew Lansley: http://www.badscience.net/2011/02/andrew-
lansley-and-his-imaginary-evidence/
NHS in thirty years: http://www.bmj.com/content/341/bmj.c3843.
full
few of them properly studied: http://www.bmj.com/content/341/
bmj.c3843.full
Kay in 2002: http://www.ncbi.nlm.nih.gov/pmc/articles/
PMC1314221/
Greener and Mannion: http://www.bmj.com/content/333/7579/
1168.full
In 1995 Coulter: http://eurpub.oxfordjournals.org/content/5/4/233.
abstract
Petchley found: http://dx.doi.org/10.1016/S0140-6736%2895%
2991805-1

'NHS Operating Framework': http://www.dh.gov.uk/en/
Managingyour organisation/Financeandplanning/
Planningframework/index.htm

Working from first principles: http://www.niesr.ac.uk/event/propper.
pdf

evidence on *fixed-price* competition: http://nedwards.posterous.
com/competition-in-healthcare

death rate from heart attacks: http://www.ncbi.nlm.nih.gov/pmc/
articles/PMC1360901/

while in the UK: http://www.bristol.ac.uk/cmpo/publications/
papers/2010/abstract242.html

to take just two things: http://dx.doi.org/10.1016/S0140-6736%2810%
2961231-7

John Appleby: http://www.bmj.com/content/342/bmj.d566.full

Why Is Evidence So Hard for Politicians?

Why is Evidence: http://www.badscience.net/2011/02/why-is-
evidence-so-hard-for-politicians/

Last week we saw: http://www.guardian.co.uk/commentisfree/2011/
feb/05/lansley-use-word-evidence

misleading static figures: http://www.bmj.com/content/342/bmj.
d566.full

Paul Burstow has kindly responded: http://www.guardian.co.uk/
society/2011/feb/08/deconstruction-of-the-nhs-bill?INTCMP=
SRCH

gap is closing so rapidly: http://www.bmj.com/content/342/bmj.
d566.full

McKee and Nolte: http://www.commonwealthfund.org/Content/
Publications/In-the-Literature/2008/Jan/Measuring-the-Health-
of-Nations--Updating-an-Earlier-Analysis.aspx

'overlooked the impact assessment': http://www.dh.gov.uk/prod_
consum_dh/groups/dh_digitalassets/documents/digitalasset/
dh_123582.pdf

the four peer-reviewed academic papers: http://www.ncbi.nlm.nih.
gov/pmc/articles/PMC1314221, http://www.bmj.com/
content/333/7579/1168.extract, http://eurpub.oxfordjournals.org/
content/5/4/233.abstract and http://linkinghub.elsevier.com/
retrieve/pii/ S0140673695918051

Politicians Can Divine Which Policy Works Best by Using Their Special Magic Politician Beam

Politicians Can Divine: http://www.badscience.net/2010/05/politicians-can-divine-which-policy-works-best-by-using-their-special-magic-politician-beam/

'Programme for Government': http://www.direct.gov.uk/prod_consum_dg/groups/dg_digitalassets/@dg/@en/documents/digitalasset/dg_187876.pdf

piloted in three cities: http://rds.homeoffice.gov.uk/rds/pdfs2/r184.pdf

drug use is estimated: http://www.tdpf.org.uk/MediaNews_Fact ResearchGuide_prisons.htm

Pornography in Hospitals

Pornography in Hospitals: http://www.badscience.net/2010/09/pornography-in-hospitals/

angry about pornography: http://www.thesun.co.uk/sol/homepage/news/3129151/Taxpayers-foot-bill-for-donors-porn-on-the-National-Health- Service-says-2020healthorg-report.html

Telegraph immediately followed suit: http://www.telegraph.co.uk/health/healthnews/7988367/ NHS-buys-porn-for-sperm-donors.html

Who said pornography was acceptable: http://www.2020health.org/research/porn.html

Hemsworth and Galloway: http://www.animalreproductionscience.com/article/0378-4320%2879%2990025-3/abstract

not present in rams: http://dx.doi.org/10.1016/0168-1591%2891%2990138-N

Mader and colleagues: http://jas.fass.org/cgi/content/abstract/59/2/294?ijkey=7e44681dab39ded34d438b6638e77eebc54cda3e&keytype2=tf_ipsecsha

Price and colleagues: http://dx.doi.org/10.1016/0168-1591%2884%2990054-6

Kerruish reported: http://dx.doi.org/10.1016/S0950-5601%2855%2980049-4

Kilgallon and Simmons: http://www.ncbi.nlm.nih.gov/pmc/articles/PMC1617155/

Zbinden and colleagues: http://beheco.oxfordjournals.org/content/15/1/137.full

Yamamoto and colleagues: http://onlinelibrary.wiley.com/doi/10.1111/
 j.1439-0272.2000.tb02877.x/abstract
impossible to ejaculate: http://humrep.oxfordjournals.org/content/
 19/9/2088.abstract

The Power of Ideas

Atheist's Guide to Christmas: http://www.amazon.co.uk/gp/product
 /0007322615?ie=UTF8&ref_=sr_1_1&s=books&qid=126095759
 7&sr=1-1&linkCode=shr&camp=3194&creative=21330&tag=
 bs0b-21

'Exams Are Getting Easier'

'Exams are Getting Easier': http://www.badscience.net/2010/08/exams-
 are-getting-easier/
'The Flynn Effect': http://en.wikipedia.org/wiki/Flynn_effect
'The Five Decade Challenge': http://www.rsc.org/images/ExamReport_
 tcm18-139067.pdf
study of just this: http://www.ons.gov.uk/about-statistics/ukcemga/
 work-areas/justice--education-and-children/changes-in-standards-
 at-gcse-and-a-level.doc
'Measuring the Mathematics Problem': http://www.engc.org.uk/
 ecukdocuments/internet/document library/Measuring the
 Mathematic Problems.pdf

Over There! An Eight-Mile-High Distraction
Made of Posh Chocolate

Over There!: http://www.badscience.net/2009/08/check-me-out-i-
 bought-some-posh-chocolate-im-political/
two review papers: http://www.food.gov.uk/news/newsarchive/2009/
 jul/organic
in terms of composition: http://www.food.gov.uk/multimedia/pdfs/
 organicreviewappendices.pdf
or health benefits: http://www.food.gov.uk/multimedia/pdfs/
 organicreviewreport.pdf
blanket right of reply: http://news.google.co.uk/news?ned=uk&hl=
 en&ncl=d0-fIEkn R_72tGMiU0uORy7OgGBTM&cf=all
Don't talk about that: http://www.badscience.net/2007/04/this-
 ageing-breadhead-guy-is-totally-angry-with-me/

pharmaceutical companies before it: http://www.badscience.
net/2008/04/cliff-richard-gloria-hunniford-carole-caplin-the-
60bn-food-supplement-industry-and-the-quantum-xrroid-dude-
refute-a-cochrane-meta-analysis/
example from its press release: http://www.soilassociation.org/News/
NewsItem/tabid/91/ smid/463/ArticleID/97/reftab/57/t/Soil-
Association-response-to-the-Food-Standards-Agency-s-Organic-
Review/Default.aspx
www.qlif.org: http://www.qlif.org/
list of 120 papers: http://orgprints.org/view/projects/eu_qlif.html
immune parameters in rat: http://orgprints.org/12653/
Salmonella Infection Level: http://orgprints.org/13728/

As Far as I Understand Thinktanks . . .
As Far as I: http://www.badscience.net/2008/06/707/

Meaningful Debates Need Clear Information
Meaningful Debates: http://www.badscience.net/2007/10/557/
in the *Independent*: http://news.independent.co.uk/uk/politics/
article3084306.ece
and the *Telegraph*: http://www.telegraph.co.uk/news/main.jhtml?xml=/
news/2007/10/15/nabortion115.xml
on Channel 4: http://www.channel4.com/news/articles/society/health/
abortion+limit+row+looming/924147
anything to declare: http://www.guardian.co.uk/science/2007/oct/15/
sciencenews.medicineandhealth
Further references:
Here is the oral evidence: www.parliament.uk/parliamentary_commit-
tees/science_and_technology_committee/scitechfm151007.cfm
Here are the memos, Prof Wyatt's are the last two in this PDF, and
one earlier one: www.parliament.uk/documents/upload/
SDAevidence.pdf

Minority Report
Minority Report: http://www.badscience.net/2007/11/minority-
retort/
published as an appendix: http://www.badscience.net/wp-content/
uploads/hc-1045-i-final-abortion-report.pdf

where they talk about me: http://www.badscience.net/2007/10/
oooooh-im-in-the-minority-report/
on 27 October an article: http://www.badscience.net/2007/10/557/

DRUGS

A Rock of Crack as Big as the Ritz

A Rock of Crack: http://www.badscience.net/2009/02/a-rock-of-
crack-as-big-as-the-ritz/

Daily Mail headline: http://www.dailymail.co.uk/health/article-
1149207/How-using-Facebook-raise-risk-cancer.html

Facebook Could Raise your Risk: http://www.nhs.uk/
news/2009/02February/Pages/Facebookhealthstudy.aspx

said the *Guardian*: http://www.guardian.co.uk/world/2009/feb/18/
taliban-british-troops-drugs

In the *Telegraph*: http://www.telegraph.co.uk/news/newstopics/
onthefrontline/4681443/British-forces-in-Afghanistan-seize-50m-
of-heroin-and-kill-20-Taliban.html

MoD press release: http://www.mod.uk/DefenceInternet/
DefenceNews/ MilitaryOperations/
HelicopterborneTroopsStrikeAtTalibansDrugIndustry.htm

chemicals and vats: http://www.metimes.com/International/2008/
05/14/converting_afghan_opium_into_heroin/3696/

2008 world report: http://www.unodc.org/documents/crop-
monitoring/Afghanistan_Opium_Survey_2008.pdf

cheap and easy: http://www.pbs.org/wgbh/pages/frontline/shows/
heroin/transform/

wholesale price fallen dramatically: http://www.unodc.org/documents/
wdr/WDR_2008/WDR2008_Statistical_Annex_Prices.pdf

The Least Surrogate Outcome

The Least Surrogate: http://www.badscience.net/2008/04/the-least-
surrogate-outcome/

aren't very informative: http://www.jstor.org/pss/3552863

or reliable: http://www.annals.org/cgi/content/full/129/12/1066?ck=nck

Drugs: Protecting Families and Communities: http://drugs.
homeoffice.gov.uk/publication-search/drug-strategy/drug-action-
plan-2008-2011?view=Binary

Public Service Agreement Delivery: http://www.hm-treasury.gov.uk/
media/B/1/pbr_csr07_psa25. pdf

Heroin on Prescription

Berridge, V., Edwards, G. (1981), *Opium and the People*,
Harmondsworth: Penguin

Caplehorn, J.R., Irwig, L., Saunders, J.B., 'Attitudes and beliefs of
staff working in methadone maintenance clinics', *Subst Use
Misuse*, 1996 Mar; 31(4): 437–52

Clark, D. (1980), 'Smack in the capital', *Time Out*, no. 51,
pp.11–13

Dole V.P., Nyswander, M. A., 'medical treatment for diacetyl
morphine (heroin) addiction', *JAMA* 1965;193:80-4

Dorn, N., Baker, O., Seddon, T. (1994), 'Paying for heroin: estimating
the financial cost of acquisitive crime committed by dependent
heroin users in England and Wales', Institute for the Study of
Drug Dependence (ISDD), London, 1994

European Centre for the Epidemiological Monitoring of AIDS, 1996,
cited in UNDCP 1997, p. 311

Follett et al., 'HIV antibody in drug-abusers in the West of
Scotland: the Edinburgh connection', *Lancet*, 1986 i, 446

Gilman, M., Pearson, G., 'Lifestyles and law enforcement' in *Policing
and Prescribing: The British System of Drug Control*, eds. Whynes,
D.K., Bean, P.T. Macmillan: London, 1991

Glossop, M., 'Verschreibung von Heroin und anderen injizierbaren
Drogen an Ahbangige aus Britischer Sicht', *Sucht* 1994; 5: 325-33
(translated in 'report of the committee of the health council of
the Netherlands', 1995)

Gossop, M., Strang, J., Connell, P., 'The response of outpatient opiate
addicts to the provision of a temporary increase in their prescribed
drugs', *Br J Psych* 1982; 141: 338-43

Gossop, M., Strang, J. (1991), 'A comparison of the withdrawal
responses of heroin and methadone addicts dutin detoxification',
Br J Psych 1991; 158: 697-9

Harding-Pink, D., 'Methadone: one person's maintenance dose is
another man's poison', *Lancet* 1996; 341.

HMSO Ministry of Health (1965), *Drug Addiction: the second report
of the inter-departmental committee*. London: HMSO.

Institute for the Study of Drug Dependence (1994), *Drug Misuse in Britain 1994*. London: ISDD.

ISDD Druglink 1996: 11i p.6

Lewis, R., 'Serious business: the global heroin economy' in Henman, A., Lewis, R., Malyon, T. (eds.), *The Big Deal*. London: Pluto Press, 1985.

Marks, J.A., 'The North Wind and the Sun', *Proc Roy Coll Phys Edin* 1991 21(3); 319-327

Maudsley, G., Williams, E., 'Inaccuracy in death certification: where are we now?', *J Public Health Med* 1996; 18: 59-66

Mott, J. 'Crime and Heroin Use' in *Policing and Prescribing: The British System of Drug Control*, eds. Whynes, D.K., Bean, P.T. Macmillan: London, 1991.

Newcombe, R. (1992), 'The reduction of drug-related harm: a conceptual framework for theory, practice, and research' in O'Hare et al (eds.), *The Reduction of Drug Related Harm*. London: Routledge, 1992

Newcombe, R., ISDD Druglink 1996: 11(i), pp.9–12

Newcombe R, Parker H., 'Heroin use and acquisitive crime in an English Community', *Br J Sociol* 1987, 38: 331–350

Payne-James, J.J., Dean, P.J., Keys, D.W. (1994), 'Drug misusers in police custody', *J R Soc Med* 1994; 87: 13–14

Plant, M.A., *Drugs in Perspective*. London: Hodder and Stoughton, 1987

Robertson et al (1986), 'Epidemic of AIDS-related virus infection among intravenous heroin users', *British Medical Journal* 292, pp.527–9.

Sobell, L.C. (1990), 'The aftermath of heresy: drinking and life events', in Miller, W.R., Greeley, J.G. (eds.), *Proceedings of the Fifth International Conference on the Treatment of Addictive Behaviours (ICTAB-5)*. Sydney: Pergamon Press.

Spear, H.B. (1969), 'The growth of heroin addiction in the UK', *British Journal of Addiction*, 64, pp. 245–55

Stewart, T., *The Heroin Users*. Guernsey: Guernsey Press, 1987

Stimson, G.V., Oppenheimer, E., Thorley, A. (1978), 'Seven year follow up of heroin addicts', *British Medical Journal*, 1978: i 11

Stimson G.V., Oppenheimer E. (1982), *Heroin Addiction*. London: Tavistock Press

United Nations Drug Control Program, *World Drug Report*.
Oxford: Oxford University Press, 1997
Vaillant, G.E., 'Centennial Address, Society for the Study of Addiction
to Alcohol and Other Drugs'. *Proc R Soc Med*, November 1984
World Health Organisation, The HIV/AIDS Pandemic, 1994
Overview, WHO/GPA/TCO/SEF/94.4, 1994

LIBEL

NMT Is Suing Dr Peter Wilmshurst. So How Trustworthy Is This Company? Let's Look at Its Website . . .

NMT is Suing: http://www.badscience.net/2010/12/nmt-are-suing-dr-
wilmshurst-so-how-trustworthy-are-they/
MIST trial was funded: http://circ.ahajournals.org/cgi/reprint/
CIRCULATIONAHA.107.727271v1
go to its website: http://www.nmtmedical.com/
outcome of MIST trial: http://circ.ahajournals.org/cgi/reprint/
CIRCULATIONAHA.107.727271v1
really was negative: http://circ.ahajournals.org/cgi/content/short/
CIRCULATIONAHA.107.727271v1
a lengthy correction: http://circ.ahajournals.org/cgi/content/
full/120/9/e71
2005 NMT report: http://www.snl.com/IRWebLinkX/GenPage.aspx?
IID=4148066&GKP=202513
restricted by GMC: http://webcache.gmc-uk.org/minutesfiles/2063.
html
fallen from $20: http://www.nasdaq.com/aspx/dynamic_charting.
aspx?selected=NMTI&symbol=NMTI

'We Are More Possible Than You Can Powerfully Imagine'

Guardian: http://www.guardian.co.uk/commentisfree/libertycentral/
2009/jul/29/simon-singh-science-chiropractic-litigation
intending to do good: http://ebn.bmj.com/cgi/content/full/10/1/4
British Medical Journal: http://www.google.co.uk/url?sa=t&source=
web&ct=res&cd=1&url=http%3A%2F%2Fwww.bmj.com%2F&ei=
VURwSp3tMMahjAfo282hBQ&usg=AFQjCNFrgA5ACj4Qwf7dYRi
vBOSsYx2Hrg&sig2=2-JerEf-4W8otKp9mO4J7A

The Times: http://www.timesonline.co.uk/tol/news/article6426195.
ece

Daily Mail: http://www.dailymail.co.uk/debate/article-1196696/Back-
cures-brave-scientist-epic-court-battle-How-Britains-libel-laws-
threatening-free-speech.html

Daily Telegraph: http://www.telegraph.co.uk/science/science-news/
5442522/Stephen-Fry-and-Ricky-Gervais-defend-science-writer-
sued-for-libel.html

Independent: http://www.independent.co.uk/life-style/health-and-
families/ features/jeremy-laurance-the-libel-laws-that-threaten-
to-stifle-scientific-debate-1744810.html

Nature: http://www.nature.com/nature/journal/v459/n7248/full/
459751a.html

Economist: http://www.economist.com/displaystory.cfm?story_id=
13809291

Times Higher Education: http://www.timeshighereducation.co.uk/
story.asp?sectioncode=26&storycode=406872&c=1

Sunday Times: http://www.timesonline.co.uk/tol/comment/
article6493564.ece

Channel 4: http://www.channel4.com/news/articles/uk/ouch+dr+
singh+hits+back/3194057

Wall Street Journal: http://online.wsj.com/article/
SB124406714025182743.html

Observer: http://www.guardian.co.uk/commentisfree/2009/may/31/
simon-singh-science

British Medical Journal: http://www.bmj.com/cgi/content/full/339/
jul08_4/b2783

international petition: http://www.senseaboutscience.org.uk/index.
php/site/project/333

Professor David Colquhorn: http://www.dcscience.net/?p=1775

BCA selectively quoting: http://layscience.net/node/598

Every stone was turned: http://www.layscience.net/node/598

Quackometer: http://www.quackometer.net/

APGaylard: http://apgaylard.wordpress.com/2009/06/18/plethora-or-
paucity-the-bca-and-bedwetting/

Gimplyblog: http://gimpyblog.wordpress.com/2009/06/18/
the-bca-have-no-evidence-that-chiropractic-can-help-with-
ear-infections/

EvidenceMatters: http://evidencematters.org/2009/06/18/british-chiropractic-association-and-the-plethora-of-evidence-for-paediatric-asthma/

Dr Petra Boynton: http://www.drpetra.co.uk/blog/?p=857

Ministry of Truth: http://www.ministryoftruth.me.uk/2009/06/18/examining-the-bcas-plethora-of-evidence/

Holfordwatch: http://holfordwatch.info/2009/06/18/british-chiropractic-association-bca-demonstrate-what-evidence-based-medicine-isnt/

Jack of Kent: http://jackofkent.blogspot.com/2009/06/bcas-worst-day-yet.html

Advertising Standards Authority: http://www.zenosblog.com/2009/07/another-asa-win-against-quackery.html

breached the guidelines: http://www.asa.org.uk/asa/adjudications/Public/TF_ADJ_46281.htm

individual chiropractors' claims: http://jdc325.wordpress.com/2009/05/14/bassett-chiropractic/

blogger Simon Perry: http://adventuresinnonsense.blogspot.com/2009/06/500-chiropractors-reported-to-trading.html

exactly the same thing: http://www.zenosblog.com/2009/06/omnibus-complaint-to-general.html

on Quackometer: http://www.quackometer.net/blog/2009/06/chiropractors-told-to-take-down-their.html

Science Is About Embracing Your Knockers

Science is About: http://www.badscience.net/2010/11/science-is-about-embracing-your-knockers/

Maria Hatzistenfanis: http://www.rodial.co.uk/about_maria/marias_profile.aspx

This is her crime: http://www.dailymail.co.uk/femail/beauty/article-1316454/Rodial-Boob-Job-gel-Cream-increases-bust-size-goes-sale-125.html

do in a checklist: http://www.badscience.net/2009/09/how-to-read-articles-about-health-by-dr-alicia-white/

The Return of Dr McKeith

The Return: http://www.badscience.net/2010/07/and-then-i-was-incompetently-libelled-by-a-litigious-millionaire/

'Gillian McKeith': http://www.badscience.net/2007/02/ms-gillian-mckeith-banned-from-calling-herself-a-doctor/
linked to from: http://gillianmckeith.info/
still linked from: http://gillianmckeith.tv/

QUACKS

The Noble and Ancient Tradition of Moron-Baiting
The Noble and Ancient: http://www.badscience.net/2010/05/the-noble-and-ancient-tradition-of-moron-baiting/
Martin Gardner died: http://www.guardian.co.uk/science/2010/may/27/martin-gardner-obituary
Aids denialism: http://www.nytimes.com/2006/03/13/business/media/13harpers.html

How Do You Regulate Wu?
How Do You: http://www.badscience.net/2010/02/how-do-you-regulate-wu/
Ying Wu: http://uk.reuters.com/article/idUKTRE61G3N420100217
babies with colic: http://adventuresinnonsense.blogspot.com/2009/11/cnhc-wishes-to-place-on-formal-record_27.html
shameful lengths over many: http://www.dcscience.net/?p=2351
appeals at the highest level: http://www.dcscience.net/?p=2485
Bachelor of Science degree: http://www.dcscience.net/Pittilo-consultation-What-is-taught.pdf

Blame Everyone But Yourselves
Blame Everyone: http://www.badscience.net/2008/07/blame-everyone-but-yourselves/
In the *Daily Telegraph*: http://www.telegraph.co.uk/news/uknews/2445080/Barbara-Nash-case-highlights-lack-of-regulation-of-nutritionists.html
advertised on yell.com: http://www.yell.com/quickclicks/SP/N/Nutritionists_and_Dieticians/berkshire/1
carries such privileges: http://www.bant.org.uk/bant/jsp/benefitsOf-Joining.faces

carries testimonials: http://64.233.183.104/search?q=cache:
f4ezlNZzh4cJ:www.barbaranash.co.uk/nutritionaltestimonials.
html+barbara+nash&hl=en&ct=clnk&cd=3&gl=uk&client=fir
efox-a

MAGIC BOXES

ADE 651: WTF?
ADE 651: http://www.badscience.net/2009/11/wtf/
New York Times: http://www.nytimes.com/2009/11/04/world/
middleeast/04sensors.html?_r=4&hp
The ADE 651: http://www.ade651.com/ade651in.html
Sandia Labs: http://www.sandia.gov/mission/homeland/
factsheets/2008/NEXESS_Factsheet_Mar08_2.pdf
tested various similar devices: http://www.justnet.org/Lists/
JUSTNET%Resources/Attachments/440/moleeval_apr02.pdf
invited the manufacturers: http://www.randi.org/site/index.php/swift-
blog/231-a-direct-specific-challenge-from-james-randi-and-the-jref.
html

After Madeleine, Why Not Bin Laden?
After Madeleine: http://www.badscience.net/2007/10/after-madeleine-
why-not-bin-laden/
'matter orientation system machine': http://www.google.co.uk/search
?q=%22matter+orientation+sy stem+machine%22
said the *Observer*: http://observer.guardian.co.uk/uk_news/story/
0,,2185358,00. html
million-dollar prize: http://www.randi.org/research/index.html
South African documentary: http://www.carteblanche.co.za/Display/
Display.asp?Id=3233
lost a child: http://www.mirror.co.uk/news/topstories/2007/10/08/
don-t-trust-the-bodyfinder-89520-19913118/

Who's Holding the Smoking Gun on Bioresonance?
Who's Holding: http://www.badscience.net/2005/11/whos-holding-
a-smoking-gun-to-bioresonance/
video is online: http://www.badscience.net/files/bioresonance.wmv

AIDS

House of Numbers

House of Numbers: http://www.badscience.net/2009/09/house-of-numbers/

Guardian Science podcast http://www.guardian.co.uk/science/blog/audio/2009/sep/21/ science-weekly-podcast-darwin-dawkins

since recanted: http://blog.newhumanist.org.uk/2009/09/was-i-conned-by-aids-denialists.html

Aids Denialism at the *Spectator*

Aids Denialism: http://www.badscience.net/2009/10/aids-denialism-at-the-spectator/

from three weeks ago: http://www.guardian.co.uk/commentis free/2009/sep/26/ben-goldacre-bad-science-aids

senior programmer appeared: http://gimpyblog.wordpress.com/2009/10/04/film-festival-endorse-aids-denialism/

Fraser Nelson: http://www.spectator.co.uk/coffeehouse/5461313/questioning-the-aids-consensus.thtml

a *Spectator* event: http://www.spectator.co.uk/shop/events/5402473/spectator-debate-a-world-without-aids.thtml

in two separate studies: http://afraf.oxfordjournals.org/cgi/content/abstract/107/427/157 and http://www.ncbi.nlm.nih.gov/pubmed/18931626

has also, oddly, flirted: http://www.spectator.co.uk/coffeehouse/5269468/the-flu-jab-choice-the-department-of-health-might-not-tell-you-about. thtml

Charles Geshekter: http://www.virusmyth.com/aids/index/cgeshekter.htm

'The Plague That Isn't': http://www.virusmyth.com/aids/index/cgeshekter.htm

emeritus professor: http://www1.imperial.ac.uk/medicine/people/b.griffin/

is quoted by virusmyth: http://www.virusmyth.com/aids/index/bgriffin.htm

overall beneficial treatment: http://www.bmj.com/cgi/content/full/324/7340/757

ELECTROSENSITIVITY

Wi-Fi Wants to Kill Your Children . . . But Alasdair Philips of Powerwatch Sells the Cure!

Wi-Fi Wants: http://www.badscience.net/2007/05/so-simple-a-child-
  could-spot-it/

Panorama programme: http://www.badscience.net/?p=414

Alasdair Philips: http://www.badscience.net/?p=241

Powerwatch: http://www.powerwatch.org.uk/

sell one to you: http://www.emfields.org/equipment/overview.asp

you can buy: http://www.emfields.org/screening/overview.asp

mesh beekeeper hat: http://www.emfields.org/screening/headnet.asp

produced 'radiation' scares: http://www.wellingtongrey.net/miscel
  lanea/archive/2007-05-27--the-truth-about-wireless-devices.
  html

Why Don't Journalists Mention the Data

Why Don't Journalists: http://www.badscience.net/2007/06/
  bmj-column-why-dont-journalists-mention-the-data/

'provocation studies' published: http://www.badscience.net/?
  p=239

POST-MODERNISM

Archie Cochrane: 'Fascist'

Archie Cochrane: http://www.badscience.net/2006/08/archie-
  cochrane-fascist/

in your summer holidays: http://www.cardiff.ac.uk/schoolsand
  divisions/divisions/insrv/libraryservices/scolar/archives/
  cochrane/biography.html

IRRATIONALITY

The Golden Arse-Beam Method

The Golden Arse-Beam: http://www.badscience.net/2011/07/the-
  golden-arse-beam-method/

One example was published: http://www.tandfonline.com/doi/full/
  10.1080/08870441003703218

subgroup analyses: http://www.badscience.
net/2011/07/2009/04/a-frankly-thin-contrivance-for-writing-on-
the-fascinating-issue-of-subgroup-analysis/
An earlier study: http://psp.sagepub.com/content/35/1/60
homeopathy: http://www.guardian.co.uk/lifeandstyle/homeopathy

Illusions of Control
Illusions of Control: http://www.badscience.net/2010/12/illusions-
of-control/
British Journal of Psychology: http://bpsoc.publisher.ingentaconnect.
com/content/bpsoc/bjp/pre-prints/bjp898
whole raft of research: http://en.wikipedia.org/wiki/Illusion_of_
control
series of studies: https://gsbapps.stanford.edu/researchpapers/
library/RP2009.pdf
connected to nothing: http://www.newyorker.com/
reporting/2008/04/21/080421fa_fact_paumgarten?currentPage=all

Empathy's Failures
Empathy's Failures: http://www.badscience.net/2010/10/empathys-
failures/
Social Psychology and Personality Science: http://spp.sagepub.com/
content/early/2010/08/24/1948550610382308.full.pdf+html
That phrasing comes: https://twitter.com/PinkZapCat/
status/488771332541546496

Blind Prejudice
Blind Prejudice: http://www.badscience.net/2010/09/blind-prejudice/
Noola Griffiths: http://www.tees.ac.uk/sections/research/social_futures/
members.cfm?griffiths=true
what woman wear: http://pom.sagepub.com/cgi/reprint/38/2/159
Goldin and Rouse: http://www.faculty.diversity.ucla.edu/search/
searchtoolkit/docs/articles/Orchestrating_Impartiality.pdf

Yeah, Well, You Can Prove Anything with Science
Yeah http://www.badscience.net/2010/07/yeah-well-you-can-prove-
anything-with-science/
Lord in 1979: http://socrates.berkeley.edu/~maccoun/ar_bias.html

Journal of Applied Social Psychology: http://www3.interscience.wiley.
com/journal/123328312/ abstract

Superstition
Superstition: http://www.badscience.net/2010/06/1693/
Psychological Science: http://pss.sagepub.com/content/early/2010/05/
27/0956797610 372631.full
Malcolm Gladwell: http://amzn.to/9QrgTe
Vaughn Bell: http://twitter.com/vaughanbell
Ed Yong: http://twitter.com/edyong209

Evidence-Based Smear Campaigns
Evidence-Based: http://www.badscience.net/2010/05/evidence-based-
smear-campaigns/
A new experiment: http://www.springerlink.com/content/
064786861r21m257/ fulltext.html

Why Cigarette Packs Matter
Why Cigarette Packs: http://www.badscience.net/2011/03/why-
cigarette-packs-matter/
government committed itself: http://www.dh.gov.uk/en/
Publicationsandstatistics/Publications/
PublicationsPolicyAndGuidance/DH_124917
put out a statement: http://www.guardian.co.uk/politics/blog/2011/
mar/09/houseofcommons-pmqs
lower-tar brands compensate: http://cancercontrol.cancer.gov/tcrb/
monographs/13/m13_3.pdf
from a million people: http://www.bmj.com/content/328/7431/72.
abstract
less likely to give up: http://cancercontrol.cancer.gov/tcrb/mono
graphs/13/m13_6.pdf
population-based surveys: http://www.ncbi.nlm.nih.gov/
pubmed/11799542
71 per cent in China: http://davidhammond.ca/Old%20Website/
Publication%20new/2010%20China%20Lights%20TC%20
%28Elton-Marshal%29.pdf
telephone digit survey: http://tobaccocontrol.bmj.com/content/10/
suppl_1/i17.abstract

five hundred smokers: http://www.ncbi.nlm.nih.gov/
    pubmed/12473429
267 adolescents: http://pediatrics.aappublications.org/cgi/content/
    abstract/114/4/e445
Careful manipulation: http://www.springerlink.com/content/
    r13q87565w885u07/?p= 56e7d0716ac94f00981ab7681baebb5
    9&pi=17
deter quitters: http://www.ncbi.nlm.nih.gov/pmc/articles/ PMC1492251
    /?tool=pubmed
perfect vehicle: http://www.springerlink.com/content/r13q87565
    w885u07/?p= 56e7d0716ac94f00981ab7681baebb59&pi=17
in four countries: http://davidhammond.ca/Old Website/
    Publication new/ITC Light Mild %28Tob Control 2008%29.pdf
street-interception survey: http://davidhammond.ca/Old Website/
    Publication new/Cigarette Pack Design %28JPH 2009%29.pdf
tobacco companies know this: http://davidhammond.ca/Old%20
    Website/Publication%20new/2010%20Malrbor%20Alchemy%20
    TC%20%28Thrasher%20et%20al%29.pdf
Marketing New Products: http://www.pmdocs.com/getallimg.asp?if
    =avpidx&DOCID=2044762173/2364
'brand imagery' studies: http://legacy.library.ucsf.edu/tid/mdb51f00
six hundred adolescents: http://www.ncbi.nlm.nih.gov/
    pubmed/1529094

All Bow Before the Mighty Power of the Nocebo Effect
All Bow Before: http://www.badscience.net/2009/11/all-bow-before-
    the-mighty-power-of-the-nocebo-effect/
looked into the evidence: http://www.badscience.net/2009/11/
    parliamentary-science-and-technology-select-committee-on-
    homeopathy-today/
Pain: http://www.painjournalonline.com/article/S0304-
    3959%2809%2900399-6/abstract
study in 2006: http://www.psychosomaticmedicine.org/cgi/content/
    abstract/68/3/478
paper in 2004: http://www.ncbi.nlm.nih.gov/pubmed/15301298
paper from 1987: http://www.ncbi.nlm.nih.gov/pubmed/3621780?
    dopt=Abstract
systematic review: http://dx.doi.org/10.1016/S1475-4916(03)00007-9

So Brilliantly You've Presented a Really Transgressive Case Through the Mainstream Media

So Brilliantly: http://www.badscience.net/2009/12/so-brillliantly-youve-presented-a-really-transgressive-case-through-the-mainstream-media/

Sun: http://www.thesun.co.uk/sol/homepage/news/2743062/23-year-nightmare-of-Rom-Houben-wrongly-diagnosed-as-comatose.html

BBC: http://news.bbc.co.uk/1/hi/world/europe/8375326.stm

Guardian: http://browse.guardian.co.uk/search?search=rom+houben&sitesearch-radio=guardian&go-guardian=Search

Telegraph: http://www.telegraph.co.uk/comment/columnists/lizhunt/6649381/Rom-Houben-and-the-human-spirit-that-would-not-be-denied.html

Der Spiegel: http://www.spiegel.de/spiegel/0,1518,662625,00.html

Australian TV news: http://news.ninemsn.com.au/world/975121/belgian-coma-man-was-just-awake-for-23-years

author of the communication: http://www.dcsf.gov.uk/research/data/uploadfiles/RR77.pdf

more recent studies: http://kslinker.com/facilitated-communicaton-since-1995.pdf

National Autistic Society says: http://www.nas.org.uk/nas/jsp/polopoly.jsp?a=3285&d=1384

position paper on FC: http://www.apa.org/about/division/cpmscientific.html

encourage you to do so: qurl.com/coma

This was barely reported: http://www.npr.org/blogs/health/2010/02/bookwriting_man_in_coma_flunks.html

BAD JOURNALISM

Asking for It

Asking for It: http://www.badscience.net/2009/07/asking-for-it/

Women who dress provocatively: http://current.com/items/90259840_women-who-dress-provocatively-more-likely-to-be-raped-claim-scientists.htm

Promiscuous men more likely: 50 http://www2.le.ac.uk/ebulletin/news/press-releases/2000-2009/ 2009/06/nparticle.2009-06-23.2976340719

blame on women: http://psychcentral.com/news/2009/06/25/
men-blame-rape-victims-more-than-women/6736.html

Jab 'as Deadly as the Cancer'

Jab 'as Deadly': http://www.badscience.net/2009/10/jabs-as-bad-as-
the-cancer/

debate at Royal Institution: http://www.timeshighereducation.co.uk/
webcast.html

'as deadly as the cancer': http://209.85.229.132/
search?q=cache:ckeH3LgaZO4J:www.express.co.uk/posts/
view/131817/Jab-as-deadly-as-the-cancer-

previous work includes: http://www.journalisted.com/lucy-
johnston

Doctor's MMR fears: http://www.express.co.uk/posts/view/112286/
Doctor-s-MMR-fears

'Linked to Phone Masts': http://www.badscience.net/category/roger-
coghill/

published a long time ago: http://www.ncbi.nlm.nih.gov/
pubmed/19501728

disappeared from *Express* website: http://www.express.co.uk/posts/
view/131817/Jab-as-deadly-as-the-cancer-

Health Warning: Exercise Makes You Fat

Health Warning: http://www.badscience.net/2009/08/health-
warning-exercise-makes-you-fat/

forty-three trials: http://mrw.interscience.wiley.com/cochrane/
clsysrev/articles/CD003817/frame.html

exercise for weight loss: http://mrw.interscience.wiley.com/cochrane/
clsysrev/articles/CD003817/frame.html

The Caveat in Paragraph Number 19

The Caveat: http://www.badscience.net/2010/10/the-caveat-in-
paragraph-number-19/

Oncological Ontology Project: http://thedailymailoncologicalontol
ogyproject.wordpress. com/

Kill Or Cure: http://kill-or-cure.heroku.com/

40 per cent: http://www.dailymail.co.uk/health/article-1317856/
Strict-diet-days-week-cuts-risk-breast-cancer-40-cent.html

the academic paper: http://www.ncbi.nlm.nih.gov/
pubmed/20921964
look at the penis: http://www.ojr.org/ojr/stories/070312ruel/
early study in 1990: http://eyetrack.poynter.org/previous.html
most recent project: http://eyetrack.poynter.org/keys_01.html

Why Don't Journalists Link to Primary Sources

Why Don't Journalists: http://www.badscience.net/2011/03/why-dont-
journalists-link-to-primary-sources/
Wind farms blamed: http://bengoldacre.posterous.com/how-far-will-
the-daily-telegraph-distort-a-st
open-access academic paper: http://www.plosone.org/article/
info%3Adoi%2F10.1371%2Fjournal.pone.0017009
end of the press release: https://www.st-andrews.ac.uk/news/
archive/2011/Title,65795,en.html
now deleted: http://www.telegraph.co.uk/earth/energy/windpower/
8382476/Wind-farms-blamed-for-stranding-of-whales.html
miserly correction: http://www.telegraph.co.uk/earth/energy/wind
power/8388273/Correction-whales-and-wind-farms.html
secret to shapely legs: http://fashion.telegraph.co.uk/news-
features/TMG8296453/Why-stilettos-are-the-secret-to-shapely-
legs.html
shapelier legs than flats: http://www.dailymail.co.uk/femail/article-
1352831/Victoria-Beckham-Stilettos-women-shapelier-legs-flats.
html
Stilettos tone up: http://www.express.co.uk/posts/view/226658/
Stilettos-tone-up-your-legs/Stilettos-tone-up-your-legs
read even the press release: http://www.hmc.edu/newsandevents/
ahn-examines-the-human-leg.html
Daily Mail: http://scienceblog.cancerresearchuk.org/2011/03/17/
no-need-to-worry-about-having-a-shower-or-drinking-water/
read the original paper: http://www.ehjournal.net/content/pdf/1476-
069X-10-18.pdf
or even the press release: http://www.creal.cat/en_noticies/view.
php?ID=85
simple distortion: http://scienceblog.cancerresearchuk.
org/2011/03/17/ no-need-to-worry-about-having-a-shower-or-
drinking-water/

A Fishy Friend, and His Friends

A Fishy Friend: http://www.badscience.net/2010/06/the-return-of-a-2bn-fishy-friend/

Fish oil helps: http://www.guardian.co.uk/science/2010/may/30/fish-oil-supplement-concentration

omega-3 fish oil pill: http://www.badscience.net/category/fish-oil/

Denis Campbell: http://www.badscience.net/2007/07/british-medical-journal-mmr-the-scare-stories-are-back/

was indeed the paper: http://www.ajcn.org/cgi/content/abstract/91/4/1060

published in full this year: http://dx.doi.org/10.1016/j.ridd.2010.01.014

estimates global sales: http://www.nutraingredients-usa.com/Consumer-Trends/Markets-Leaders-in-global-brain-food-sales

reproduced below in full: http://www.independent.co.uk/life-style/health-and-families/ features/jeremy-laurance-dr-goldacre-doesnt-make-everything-better-1994017.html

MMR: The Scare Stories Are Back

MMR: http://www.badscience.net/2007/07/british-medical-journal-mmr-the-scare-stories-are-back/

Prevention Is Better Than Cure When It Comes to Health Scares

suspected death: http://www.bbc.co.uk/news/uk-wales-22385218

vaccine caused autism: http://www.nhs.uk/Conditions/vaccinations/Pages/mmr-vaccine.aspx

dozens of vaccine scares: http://www.ft.com/cms/s/0/0f90ac7a-a67c-11e2-885b-00144feabdc0.html

Antivaccination campaigners: http://www.ft.com/cms/s/2/a2f9f1a6-50e3-11e0-8931-00144feab49a.html

part of a plot: http://www.ft.com/cms/s/0/2a9e6704-b0d2-11e2-9f24-00144feabdc0.html

Dodgy Academic PR

Dodgy Academic PR: http://www.badscience.net/2009/05/dodgy-academic-pr/

Annals of Internal Medicine: http://www.annals.org/cgi/content/
abstract/150/9/613

Suicide
Suicide: http://www.badscience.net/2009/03/suicide/
it has been shown: http://www.ncbi.nlm.nih.gov/pubmed/11630757
repeatedly: http://dx.doi.org/10.1016/0049-089X%2891%2990016-V
increased by 17 per cent: http://www.bmj.com/cgi/content/
full/318/7189/972?view=long &pmid=10195966
a significant decrease: http://www.springerlink.com/content/
98rw3lycjnkgg9a3/

Roger Coghill and the Aids Test
Roger Coghill: http://www.badscience.net/2008/06/roger-coghill-fails-
the-aids-test/
linked to phone masts: http://www.express.co.uk/posts/view/49330/
Suicides-linked-to-phone-masts-
evidence of a possible link: http://info.cancerresearchuk.org/healthy
living/cancercontroversies/howdoweknow/
Broad Street pump: http://en.wikipedia.org/wiki/John_Snow_
(physician)
a 'stakeholder' group: http://www.rkpartnership.co.uk/sage/
specialises in mediation: http://www.rkpartnership.co.uk/whatwedo.
php
their last document: http://www.rkpartnership.co.uk/sage/Public/
SAGE first interim assessment.pdf
angel investment: http://www.walesonline.co.uk/business-in-wales/
business-news/2008/02/02/business-angel-supplements-for-
launch-91466-20426842/
'Asphalia': http://www.asphalia.co.uk/
visited his website: http://www.galonja.co.uk/
protection equipment: http://www.galonja.co.uk/galonja_shop/
product.asp?g_s_n=crlshop&g_u_no=0&g_u_nam=&g_
tim=&pid=88&v_ det=1&full=1&c_id=0
Acousticom: http://www.galonja.co.uk/galonja_shop/product.
asp?g_s_n=crlshop&g_u_no=0&g_u_nam=&g_tim=&pid=104&v_
det=1&full=1&c_id=0

makes wine taste nicer:http://www.galonja.co.uk/galonja_shop/
   product.asp?g_s_n=crlshop&g_u_no=0&g_u_nam=&g_tim=
   &pid=98&v_ det=1&full=1&c_id=0
'Mood Maker': http://www.galonja.co.uk/galonja_shop/product.
   asp?g_s_ n=crlshop&g_u_no=0&g_u_nam=&g_tim=&pid=97&v_
   det=1&full=1&c_id=0
Electrohealing: http://www.galonja.co.uk/galonja_shop/product.
   asp?g_s_ n=crlshop&g_u_no=0&g_u_nam=&g_tim=&pid=46&v_
   det=1&full=1&c_id=0
Atlantis: http://www.galonja.co.uk/galonja_shop/product.asp?g_s_
   n=crlshop&g_u_no=0&g_u_nam=&g_tim=&pid=59&v_ det=
   1&full =1&c_id=0
The Aids test: http://www.badscience.net/2007/12/aids-quackery-
   international-tour/, http://www.badscience.net/2007/09/homeop-
   athy gives-you-aids/ and http://www.badscience.net/2007/02/
   money-is-not-the-only-barrier-to-aids-patients-getting-hold-of-
   drugs/
Coghill on Aids: http://web.archive.org/web/20041013110533/http://
   www.cogreslab.co.uk/aids.htm

BRAINIAC

Ka-Boom! Science! COOL!!?!
Ka-Boom!: 831 http://www.badscience.net/2006/07/ka-boom/

Who's the Daddy?
Who's the Daddy?: http://www.badscience.net/2006/07/whos-the-
   daddy/

STUFF

Here's My . . . Foreword to the Romney, Hythe and Dymchurch Railway Guidebook
Here's my: http://www.badscience.net/2013/12/heres-my-intro-to-
   the-romney-hythe-and-dimchurch-railway-guidebook/
narrow gauge railway: http://www.rhdr.org.uk/
sound mirrors: http://en.wikipedia.org/wiki/Acoustic_mirror

How I Stalked My Girlfriend

How I Stalked: http://www.badscience.net/2006/02/how-i-stalked-my-girlfriend/

EARLY SNARKS

Staying Beautiful Is Easy to Do

Staying Beautiful: http://www.badscience.net/2003/05/staying-beautiful-is-easy-to-do/

Because You're Worth It

Because You're: http://www.badscience.net/2003/11/because-youre-worth-it/

More Than Water?

More than Water?: http://www.badscience.net/2004/01/more-than-water/

'Nanniebots' to Catch Paedophiles

'Nanniebots': http://www.badscience.net/2004/03/nanniebots-to-catch-paedophiles/
New Scientist's chat with Nanniebot: http://www.tinyurl.com/2y55h
talk to it online: http://www.tinyurl.com/2osgo

Nanniebots and Neverland

Nanniebots and Neverland: http://www.badscience.net/2004/04/nanniebots-and-neverland/
making false claims: tinyurl.com/3gfxv
modified the device to stream shows: tinyurl.com/38wmx
posting did state: tinyurl.com/2jg3p
chatnannies.com: http://chatnannies.com/

Artificial Intelligence

Artificial Intelligence: http://www.badscience.net/2004/06/artificial-intransigence/

BOOKENDS

Be Very Afraid: The Bad Science Manifesto

Be Very Afraid: http://www.badscience.net/2003/04/be-very-afraid-the-bad-science-manifesto/

What Eight Years of Writing the Bad Science Column Has Taught Me

finish a book: http://www.badscience.net/books/the-drug-pushers/

what I've learned: http://www.badscience.net/2010/12/the-year-in-nonsense-2/ and http://www.badscience.net/2009/12/the-year-in-nonsense/

in eight years: http://www.badscience.net/2008/12/the-year-in-bad-science-2/

writing this column: http://www.badscience.net/2007/12/the-year-in-bad-science-2007/ and http://www.badscience.net/2006/12/the-year-in-bad-science/

Alternative therapists: http://www.badscience.net/category/complementary-medicine/

great teaching tool: http://www.badscience.net/2007/11/a-kind-of-magic/

medicines regulators: http://www.badscience.net/2011/02/pretending-that-evidence-is-difficult-and-complicated/

universities: http://www.badscience.net/2010/02/how-do-you-regulate-wu/

science and evidence in culture: http://www.badscience.net/2008/09/the-medicalisation-of-everyday-life/

their libel cases: http://www.badscience.net/category/libel/

comedy factory: http://www.badscience.net/2008/08/bill-nelson-wins-the-internet/ and http://www.badscience.net/2007/05/the-amazing-qlink-science-pedant/

how the world works: http://www.guardian.co.uk/commentisfree/2011/jul/08/bad-science-effective-things-silly-places

the placebo effect: http://www.badscience.net/category/placebo/

misled by heuristics: http://www.badscience.net/category/irrationality-research/

thrills, and power: http://www.badscience.net/2011/03/why-cigarette-packs-matter/

Pharmaceutical companies: http://www.badscience.net/category/big-pharma/

still won't publish all: http://www.badscience.net/category/publication-bias/

we tolerate it: http://www.badscience.net/2011/03/when-regulation-is-opaque-trust-is-all-you-have/

Journalists: http://www.badscience.net/category/media/

can mislead the public: http://www.badscience.net/2010/10/the-caveat-in-paragraph-number-19/

the methods and techniques: http://www.badscience.net/2011/10/new-edition-of-testing-treatments-best-lay-text-on-evidence-based-medicine/

Politicians misuse evidence: http://www.badscience.net/category/politics/

and distort it: http://www.badscience.net/2011/04/id-expect-this-from-ukip-or-the-daily-mail-not-from-a-government-leaflet/

to shameful degrees: http://www.badscience.net/2011/02/andrew-lansley-and-his-imaginary-evidence/ and http://www.badscience.net/2011/02/why-is-evidence-so-hard-for-politicians/

trials of policies: http://www.badscience.net/2011/05/we-should-so-blatantly-do-more-randomised-trials-on-policy/

if they achieve: http://www.badscience.net/2010/05/politicians-can-divine-which-policy-works-best-by-using-their-special-magic-politician-beam/

no honourable excuse: http://www.badscience.net/2009/09/blue-print-fail/

the fairest tests: http://www.badscience.net/category/evidence-based-policy/

Real scientists: http://www.badscience.net/2011/11/why-wont-professor-greenfield-publish-this-theory-in-a-scientific-journal/

clear line between the results: http://www.guardian.co.uk/science/2011/jul/29/duchennes-muscular-dystrophy-surrogate-outcomes

nerds are more powerful: http://www.badscience.net/2011/06/kids-who-spot-bullshit-and-the-adults-who-get-upset-about-it/

best teaching gimmick: http://www.ted.com/talks/ben_goldacre_battling_bad_science.html

for explaining: http://www.guardian.co.uk/commentisfree/2011/
   oct/28/bad-science-diy-data-analysis?INTCMP=SRCH
how good: http://www.guardian.co.uk/commentisfree/2011/sep/16/
   bad-science-dodgy-stats
science works: http://www.guardian.co.uk/commentisfree/2011/
   sep/09/bad-science-research-error

List of Illustrations

Index

Index

Index

Index